CLASS TEST EDITION

DATA, EQUATIONS, AND GRAPHS
Making Sense of Elementary Algebra

Elaine Kasimatis
California State University, Sacramento

Judith Kysh
University of California at Davis

Tom Sallee
University of California at Davis

Brian Hoey
Christian Brothers High School

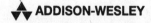 **ADDISON-WESLEY**

An imprint of Addison Wesley Longman, Inc.

Reading, Massachusetts • Menlo Park, California • New York • Harlow, England
Don Mills, Ontario • Sydney • Mexico City • Madrid • Amsterdam

Reproduced by Addison Wesley Longman from camera-ready copy supplied by the authors.

Copyright © 1997 Addison Wesley Longman.

ISBN   0-201-76800-3

1 2 3 4 5 6 7 8 9 10 CRS 00999897

Table of Contents

Chapter 6 The Amusement Park: Problems AP-1 through 100
                  Graphing and Systems of Equations

We return to graphing more complicated equations and investigate linear and non-linear systems, emphasizing the relationship between the graphs of systems and their common solution(s). In particular, you'll see that the zero-product property and solutions of quadratic equations are tightly linked to the graphs of the equations. The emphases are on understanding the physical interpretation of a graph and solving systems of linear equations graphically and algebraically.

Chapter 7 The Buckled Railroad Track: Problems RT-1 through 108
                  Words to Diagrams to Equations

In Chapter 3, you learned how to translate verbal descriptions to equations. Here the emphasis becomes translating verbal descriptions to diagrams and then to equations. Quadratic equations and square roots arise naturally from problems written using the Pythagorean theorem.

Chapter 8 The Grazing Goat: Area and Subproblems   Problems GG-1 through 95

We look more closely at the concept of area and use it to introduce the concept of subproblems in a visual way. You will use the idea of subproblems to simplify many problems, including rational expressions. The Grazing Goat Problem is a good example of a problem that includes several subproblems.

Chapter 9 The Burning Candle: More Ratios and Slope   Problems RB-1 through 88

We use similar right triangles to develop the notion of the slope of a line and to write equations, this time from a graph. You'll return to the Burning Candle Problem to develop an algebraic solution through writing and solving equations.

Chapter 10 The Election Poster: Problems EP-1 through 93
                   More About Quadratic and Linear Equations

So far, you have solved quadratic equations by Guess and Check, by graphing, and by factoring. In this chapter we give the quadratic formula as a fact to permit the solution of all quadratic equations. We designed the chapter's problems to tie together all the ideas developed throughout the course. In particular, we extend understanding of linear relationships by examining lines of best fit for data, including data for a problem similar to the Predicting Book Costs problem.

Preface

The mathematics course which follows is an adaptation of *College Preparatory Mathematics: Change from Within, Math 1 (Algebra 1)*. It is quite different from a traditional Elementary Algebra course and it reflects the experiences, vision, and philosophy of the teachers and university professors who have developed and taught CPM Math 1 at the secondary and college levels. We are not alone in wanting change. In the last decade, virtually every major document on improving mathematics instruction has called for a restructuring of the entire mathematics curriculum with more emphasis on understanding and less emphasis on routine drill. Most recently, the American Mathematical Association of Two-Year Colleges (AMATYC) has issued *Crossroads in Mathematics: Standards for Introductory College Mathematics before Calculus*, which parallels the *Curriculum and Evaluation Standards* of the National Council of Teachers of Mathematics (NCTM), one of the early guides of the development of the CPM Program. We are in an age when the mathematics which students need to know is changing dramatically -- relatively inexpensive calculators now on the market can do all of the routine manipulations which have historically made up the bulk of algebra instruction. So there is an urgent need to rethink what Elementary Algebra should be, and this course represents our efforts.

Many students who are taking algebra in college have taken algebra in high school, but may not have understood it or been able to solve problems. We believe that this course will provide what is needed for those students: better understanding of algebra through concept development, improved problem solving abilities, demystification of mathematics, increased sense of mathematical power through a focus of learning that is shifted to the student, and group methodology that encourages interaction and communication.

Note to the Student

You may be taking elementary algebra for a number of reasons, perhaps to prepare for a college-level course, or a technical field, which requires you to **use** the mathematics you have learned. The skills you learn in this course will be useful to you not only in other mathematics courses, but also in many others such as chemistry, economics, psychology, and zoology. For this reason,

YOUR AIM FOR THIS COURSE SHOULD BE UNDERSTANDING.

Learning for Understanding

Only you will know if you have understood an idea. We all know how easy it is to memorize something and even do well on a test without having any real idea about what is going on. If you settle for just performing well, you are cheating yourself, and you will likely have difficulty completing the course or using what you have learned in another environment.

We have done our best to design a course to make it comfortable for you to learn <u>and</u> to understand what you have learned. However, learning takes hard work. If you want to do well in this course and get a good start in college, there are four things you will have to do:

> - Make **understanding** the mathematics your highest goal in this course.
> - Attend class regularly.
> - Discuss questions with your group.
> - Keep up with the classwork and the homework.

On your way to understanding the mathematics you are learning, this course will help you become better at three important learning skills -- asking questions, solving problems, and explaining your reasoning. In order to ask questions, solve problems, and explain your reasoning, you need to have something to think and ask and talk about. That's why this course is built around problems, or, more specifically, it's built around **students doing problems**. You also need someone to ask and talk to. That's why the course is built around **students working together in groups**.

Working in Groups

If you want to understand the mathematics you are doing, you will need to be willing to spend time thinking and trying out alternative approaches. Often you won't be able to come up with a correct answer on the first try, so you need to be willing to persist. At this level there are generally several ways to think about a topic, and you should try to see more than one of them. Working in groups provides a natural opportunity to see and hear several approaches to one problem. That is one reason we want you to work in groups.

A second reason for working in groups is that most job situations today demand that employees work in groups, discussing ideas, listening, and testing. Colleagues must be able to combine the good parts of one person's idea with the good parts of someone else's idea to get a solution. Many of the problems in this book will ask you to discuss your ideas with your group and to listen to other people's ideas. This is an important practice to learn, so do not skip over that part of the assignment.

A third and probably most important reason for learning to work collaboratively is that we want the mathematics to make sense to **you**. Mathematical techniques should not seem random. That is why the problems in this book have been structured so that you and your group, with support from your instructor, can construct, and therefore understand, much more mathematics than you could from being given a rule and assigned a bunch of exercises which all look the same.

Working effectively in groups doesn't just "happen." It takes the conscious efforts of every student and the instructor. Here are some guidelines to help make your experiences in group work as rewarding as possible:

Guidelines for Working in Groups

1. You are responsible for your own behavior.
   (No student has the right to interfere with another student's right to learn.)

2. You must try to help anyone in your group who asks.
   (But don't give the answer or tell someone how to work a problem unless you are asked.)

3. You may ask the instructor for help only when all of the members of your group have the same question.

4. You must use a "group voice" that only members of your group can hear.
   (The volume of each student's voice should remain reasonable and within the hearing range of his/her group only.)

Asking Questions and Doing Homework

No one expects basketball players to become good (or even decent) if they just watch others play. All players have to practice diligently. But they also need to know what to practice. In mathematics, you might get stuck sometimes and not know what to do and have no one to ask for help. In such a situation, write down what you do know about the problem and write down what you tried. Then figure out what your question is and write it down, too. When you get a chance, share your notes and questions with your group members (and your instructor) so they can see how you tried to solve the problem.

Doing this kind of work is what we mean by doing homework. In fact, the homework exercises which you can easily answer do not help you learn anything new; they are just skill maintenance. Getting stuck on a problem is a **big opportunity to learn** something. Analyze the problem to find out just what the hard part is. And then when you do find out how to do the problem, ask yourself, "What was it about that problem that made it so difficult?" Answering that question will get you ready to handle the next difficult problem without getting stuck, and you will have learned something.

Strategies for Solving Problems

Much of what you learn in this course will not be brand new, but will build on mathematics which you already know. A very useful way to help yourself learn is to use such problem solving strategies as

GUESS AND CHECK, LOOK FOR A PATTERN, ORGANIZE A TABLE,
USE MANIPULATIVES, WRITE AN EQUATION, DRAW A DIAGRAM OR GRAPH,
WORK BACKWARDS, FIND A SUB-PROBLEM, or FIND AN EASIER RELATED PROBLEM.

These strategies are not only useful for solving problems, but for **learning** as well. Throughout this course you will continue to build your skills at using a variety of problem solving strategies to progress from what you know toward the solution you seek.

Three major goals for this course are for you to become better at asking questions, at solving problems, and at explaining your reasoning, all of which aim at developing understanding of mathematics. On your way to achieving these goals, you will be learning how to take responsibility for your learning. One of your responsibilities will be to seek extra practice or assistance outside of the class to help you increase your confidence level. You might exchange phone numbers with other students in the class so you can discuss questions outside of class, or talk with your instructor about extra resources for help. With patience and persistence, your efforts will pay off!

Elaine Kasimatis
Cindy Erickson
Roberta Gehrmann

Chapter 1

OPENINGS:
Data Organization

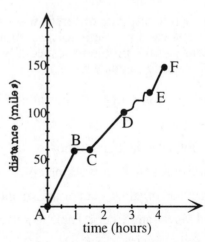

CHAPTER 1

OPENINGS:
DATA ORGANIZATION

$$1+1-(1\div 1)(1) \qquad \frac{2^2}{2+2-2}$$

$$(3\text{-}3)(3\cdot 3\cdot 3)$$

$$6+6\text{-}6+6\div 6$$

$$4+(4-4)(4+4)$$

$$(5)5\text{-}5(5+5)$$

$$7(7+7)\text{-}7+7$$

$$(8+8)(8+8\text{-}8)$$

$$9\div 9-9+(9-9)$$

In this chapter, you will investigate different ways to organize data. Working collaboratively in groups, you will also explore several problem-solving strategies as you develop and review your graphing and calculator skills. Many of the problems, especially those that involve very large or very small numbers, require the use of a scientific calculator.

Throughout the chapter, you will get better at solving problems, asking questions, and explaining your reasoning.

In this chapter you will have the opportunity to:

- establish routines for working collaboratively to solve problems;
- collect and organize data or information;
- use a Guess and Check Table to solve word problems;
- practice computing with integers;
- use the meaning of integer exponents to rewrite exponential expressions;
- read and interpret graphs;
- use your calculator more efficiently; and
- calculate simple probabilities.

CHAPTER CONTENTS

1.1 BUILDING CORRALS: WORKING COLLABORATIVELY

OP-1.   **Building Corrals** Many people find working with others helpful when trying to solve problems in mathematics. Working collaboratively can also be a challenge. Building Corrals is a "cooperative logic" problem which requires your group to work together--each of you will be given only part of the information needed to solve the problem.

>   Students who attend the University of California at Davis can board their horses at the Campus Equestrian Center. This year the Equestrian Center is home to three student horses, which until now have been boarded together in a large rectangular corral. However, the three horses have been fighting recently, and need to be separated from each other.

>   Cole, who works at the Equestrian Center, decided to subdivide the large rectangular corral to form three new corrals. Use the information on the clue cards to figure out how the new corrals could have been arranged.

            Each member of your group will receive a different clue card. Share your clue by reading it aloud, but do not show your clue card to other group members. After your group has used the clue cards to solve the problem, answer each of the following questions.

            a)   How could Cole have arranged the corrals? Sketch your solution, and then write your response in complete sentences.

            b)   Are there other solutions to the problem? If so, find them. If not, explain why not.

OP-2.   We start off the course focusing on problem solving and cooperative learning, skills that apply in many college courses and technical fields.

            Read the "Guidelines for Working in Groups" in the Note to the Student.

            a)   Think about the following questions on your own:
                  • Which guideline will be easiest for you personally to follow?
                  • Which guideline will be the most difficult?
                  • What previous experiences, if any, have you had working in groups of four?
                  • What types of things could you do to contribute to positive and productive cooperative learning experiences throughout the course? Another way to think of this is: What are the attributes of someone you would like to work with in a group?

            b)   Now discuss your responses in part (a) with your group. Take turns sharing your personal experiences and commitments.

OP-3.   Now that your group has solved the Building Corrals problem, think about how you went about finding an answer. Discuss the process your group used: How did you decide which clue to use first? Then what did you do? Write a brief outline of your group's approach to solving the problem. Be ready to explain the way your group solved the Building Corrals problem to the rest of the class.

OP-4. Write a set of clues for a new Building Corrals problem. Pass them to another group when the class is ready.

OP-5. ◊◊**Diamond Problems**

If you know the numbers in the west and east diamonds, can you find the numbers that go in the north and south?

Look at these completed diamonds and see if you can discover the pattern. How are the north and south diamonds related to the east and west diamonds?

Copy these Diamond Problems and use the pattern you found to complete each of them.

a)

b)

c)

d)

e)

f)

g)

h)

i)

j)

k)

l)

OP-6.   Do you know what multiplication means? You probably can easily recite multiplication facts, but you may not be as comfortable explaining to someone what it means to multiply two numbers.

To demonstrate how well you understand something, you must do more than simply use symbols and get correct answers. You must also be able to explain your thoughts in words and in pictures. The following triangle illustrates these three ways to make sense of a mathematical concept:

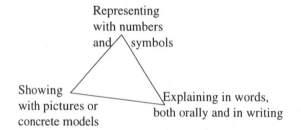

Suppose you are a volunteer in an elementary school classroom. Max, a third grader who is learning about multiplication, asks you to explain what $3 \cdot 4$ means. He says that he knows the answer is 12, but he doesn't understand why.

a)   In complete sentences, describe how you could explain verbally to Max what "3 times 4" means.

b)   Draw a picture that you could use to show Max what "3 times 4" means.

OP-7.   Copy and complete each Diamond Problem.

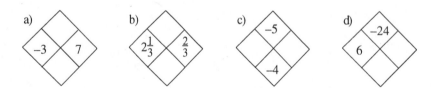

OP-8.   ⫰ **Your Mathematical Autobiography** Think about the following questions and then describe your math background in several paragraphs. What are your mathematical strengths? What are your mathematical weaknesses? Include both formal and informal experiences and training.

OP-9.   **Keeping a Notebook** You will need to keep an organized notebook for this course. Here is one method of keeping a notebook. These guidelines have worked for some students, but you should follow a system that works for you.

- The notebook should be a sturdy three-ring loose-leaf binder with a hard cover.

- The binder should have dividers to separate it into sections:
  Homework/Class work
  Notes
  Tests and Quizzes, with Corrections
  Tool Kits
  Chapter Summaries
  Supplies (blank paper, graph paper, resource pages, etc.)

- You will need to bring to class every day:
  A scientific calculator
  Lined, punched notebook paper
  Graph paper
  A ruler marked with centimeters and inches
  Pencils (having several colors is helpful)
  Erasers

Your notebook will be a big asset for learning in this course and will be your chief aid in studying for tests, so good organization is essential. Don't wait two weeks to start organizing–it will be much more efficient if you start NOW!

If you have not already done so, get a suitable binder and start organizing your notebook.

1.2 THE FIVE-DIGIT PROBLEM

OP-10.   ✋ **The Five-Digit Problem** In this problem, you will practice working with the order of operations in a collaborative setting. You'll find a record sheet for the Five-Digit Problem in your Resource Pages for this chapter, Section 1.9.

The problem is to generate the integers from 0 to 20 by using any **one** digit five times along with any mathematical symbols you like (for example: $+$, $-$, $\times$, $\div$, ...). The mathematical operations you use must be "legal."

Here are two examples:

> Suppose the class chooses to use the digit "4." Then we can use the "4" five times with division and subtraction to obtain $4 - \dfrac{4}{4} - \dfrac{4}{4}$.
>
> This expression would be recorded on the **2 line** of your paper because $4 - \dfrac{4}{4} - \dfrac{4}{4} = 2$.
>
> The expression $4 + [4 + (4 \cdot 4)] \div 4$ would be recorded on the **9 line**.

Five-Digit Problem	
0	_____
1	_____
2	$4 - \dfrac{4}{4} - \dfrac{4}{4}$
3	_____
↓	
9	$4 + [4 + (4 \cdot 4)] \div 4$
10	_____

a)   Complete ten of the five-digit number problems by writing a correct expression next to the corresponding number on your record sheet. Use a scientific calculator to check your answers.

b)   Compare your solutions with those of other group members. Record their results on your record sheet.

OP-11.   Your instructor will post a Five-Digit Problem Class Record Sheet.

a)   Choose a member of your group to record your solutions on the large chart.

b)   Now compare your expressions with those from the rest of the class. Which line has the most different expressions? Which has the fewest? Is there any number for which it is impossible to create a five-digit expression? Write several sentences about your observations.

Problems that are especially important are marked with a ⚷ symbol. They are designed either to help you develop understanding of mathematical ideas or processes, or to help you consolidate understanding. Pay careful attention to these problems, and be sure to revise your work when necessary. Many of these problems will give you ideas for Tool Kit entries.

OP-12. **Algebra Tool Kits** An Algebra Tool Kit is a useful way to keep track of methods and topics you would like to be able to find easily. The idea is to have at your fingertips a resource for figuring out "what to do when you don't remember what to do." The Algebra Tool Kit is for your own use, so make it something **you** understand. Directly copying a definition or explanation from the book will not be helpful if it does not make sense to you.

On the Tool Kit sheet provided by your instructor, you will see several topics. Some of these topics should be familiar to you. Fill in any information or examples that will be useful to you in understanding and remembering the topic of **order of operations**. You will fill in the other boxes as we reach those topics in the next few chapters.

OP-13. ▤ **Area and Perimeter**

a) Arrange nine congruent squares so that they touch along entire sides but don't overlap; then draw the arrangement on a piece of centimeter grid graph paper from the resource pages at the end of the chapter.

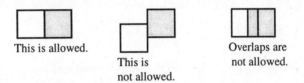

This is allowed.　　This is not allowed.　　Overlaps are not allowed.

b) Make at least five more different configurations (six in all). Each configuration should use all nine squares. Draw each arrangement on graph paper, and indicate the area and perimeter of each configuration next to its drawing.

c) What are the largest and smallest perimeters you found? What are the largest and smallest areas you found? Did any of the figures have an odd perimeter?

d) Compare your configurations with others in your group. Are they the same? What conclusions can you make about the perimeters and areas of your group's configurations? Write your observations using complete sentences.

OP-14. **$$ Dollar Bill Facts** According to information from the US. Mint, 454 US one-dollar bills weigh about one pound. Use this information to answer each of the following questions.

a) What does a single one-dollar bill weigh?

b) How much do 1 billion one-dollar bills weigh?

OP-15. Imagine you are a volunteer in a fifth grade class. Molly is starting to learn about exponents and asks you, "Why does $2^3 = 8$?" How would you respond? Write your answer in a complete sentence.

OP-16. Reread the Note to the Student, focusing on the goals. Write a paragraph that outlines your goals upon completion of this course. Include information about how this course will assist you in fulfilling your goals. Be prepared to share your paragraph with your group.

OP-17.   Copy and solve these Diamond Problems:

OP-18.   Describe in a complete sentence how the north and south diamonds in a Diamond Problem are related to the east and west diamonds.

OP-19.   Complete each of the following computations without using a calculator. Write your solutions as complete statements: $(-12)(-3) = 36$.

a) $-15 + 7$

b) $8 - (-21)$

c) $(3)(-9)$

d) $-9 + (-13)$

f) $-62.04 \div (-7.52)$

e) $-50 - 30$

g) $[15 \div (-3)] + 12$

h) $(-3 - 4) \div 7$

i) $(-6) \cdot 3^2 + (-14)$

j) $[28 - (-2)] - 12$

k) $[-3 + (-2)](-5)$

l) $(-6) \cdot 4 \div (-3)$

OP-20.   Use a calculator to check your work in problem OP-19. To enter a negative number, in most scientific calculators, you must use the change sign $\boxed{+/-}$ key, **not** the subtraction $\boxed{-}$ key. For example, to enter -10, first enter "10" and then press $\boxed{+/-}$.

OP-21.   a)   Which of the calculations in problem OP-19 could be done easily without the aid of a calculator?

b)   Which of the calculations in problem OP-19 were difficult or time-consuming to do without a calculator?

1.3 USING GUESS AND CHECK TABLES TO SOLVE PROBLEMS

OP-22.   **Using a Guess and Check Table to Solve a Problem** You have done it before. You've guessed what an answer is, and then used clues to figure out if you were right or wrong. In algebra, this is a major idea. It's major because it is a first step to mastering the power of algebra and solving problems.

Guessing in a *systematic* way is the key to making the transition from trial and error to algebraic methods. So, the examples you work and copy in class should be well organized and readable. The organization is important, because it shows how you checked each guess. This will make it easier to make the transition to more abstract mathematics later. Just as with the following examples, be neat and clear in your work. It may not be acceptable or helpful otherwise.

Copy the following example in your notebook.

Example The length of a rectangle is three centimeters more than twice the width. The perimeter is 45 centimeters. Use a Guess and Check Table to find how long and how wide the rectangle is.

STEP 1   Start a table. Label the first column as your "Guess" along with a description of what you are guessing, as in the table below. Why is the width a reasonable thing to guess?

Guess Width	

STEP 2   Make a guess--any guess--for the width. It does not need to be a "good" guess; any number will do. The idea is to get started.

Guess Width	
10	

Test your guess by writing down all of the steps you take to check it. These steps are your column descriptions. *It is essential that you carefully identify and label each column, including the "Guess" and the "Check" columns.*

STEP 3   Calculate the length of the rectangle.

Guess Width	Length	
10	23	

[PROBLEM CONTINUED ON NEXT PAGE]

OP-22. continued

STEP 4   Find the perimeter, and write out your calculations in the next column. Remember, rectangles
         have two "width" sides and two "length" sides.

Guess Width	Length	Perimeter calculations
10	23	$2 \cdot 10 + 2 \cdot 23 = 66$

STEP 5   Check the perimeter against 45, and identify it as correct, too high, or too low.

Guess Width	Length	Perimeter calculations	Check Perimeter = 45 ?
10	23	$2 \cdot 10 + 2 \cdot 23 = 66$	too high

STEP 6   Go back to Step 2 and make a new guess.

Guess Width	Length	Perimeter calculations	Check Perimeter = 45 ?
10	23	$2 \cdot 10 + 2 \cdot 23 = 66$	too high
5			

STEP 7   Follow Steps 3 through 5 using your new guess.

Guess Width	Length	Perimeter calculations	Check Perimeter = 45 ?
10	23	$2 \cdot 10 + 2 \cdot 23 = 66$	too high
5	13	$2 \cdot 5 + 2 \cdot 13 = 36$	too low

         Repeat this process until you find the solution. Use the clues from the "Check" column to
         make better and better guesses.

STEP 8   Write your answer to the problem in a complete sentence. For example,

                 "The width is 6.5 centimeters and the length is 16 centimeters."

Solve each of the following problems by using a Guess and Check Table as modeled in OP-22.
Write each solution in a complete sentence.

OP-23.   One number is five more than a second number. The product of the numbers is 3,300. Find
         the two numbers.

OP-24.   The perimeter of a triangle is 76 centimeters. The second side is twice as long as the first
         side. The third side is four centimeters shorter than the second side. How long is each side?

OP-25.   Hector is now 28 years old and Henry is eight years old. In how many years will Hector be twice as old as Henry?

OP-26.   Duke cleared the cash register of nickels and dimes. There were twenty coins in all, and the total value of the coins was $1.35. How many of each type of coin were in the cash register?

OP-27.   a)   A simple four-function calculator has keys for the basic four math operations:
$\boxed{+}$, $\boxed{-}$, $\boxed{\times}$, and $\boxed{\div}$.
Describe a way to calculate 2^3 using only the multiplication $\boxed{\times}$ and equals $\boxed{=}$ keys.

b)   To compute 2^3 quickly with a scientific calculator, use the $\boxed{y^x}$ key:
press $\boxed{2}$ $\boxed{y^x}$ $\boxed{3}$ $\boxed{=}$,
(Some calculators call it the $\boxed{x^y}$ key, but it works the same way.)
The $\boxed{y^x}$ key can be used to calculate other exponential values, too. Show how you could use it to calculate 3^6, and then write out the keystrokes as shown above.

OP-28.   Joe has never worked with integers before. Explain to him your rules for doing each of the following problems. Answer in complete sentences.

$-4 + (-2)$ $\qquad\qquad$ $3 + (-7)$ $\qquad\qquad$ $2 - 5$ $\qquad\qquad$ $3 - (-2)$

OP-29.   **Calculator Tool Kits** A Calculator Tool Kit, like an Algebra Tool Kit, can be a useful way to keep at your fingertips a set of instructions for using your calculator. Even the manual that came with your calculator can be difficult to understand -- so, write the entries in your Tool Kit carefully, so that they **make sense** to **you**!

On the Calculator Tool Kit from the resource pages at the end of the chapter, you will see several calculator keys.

a)   Write the keystrokes or examples that will be useful to you in understanding and remembering how to use the $\boxed{y^x}$ key on **your** calculator.

b)   Add helpful notes about the change sign $\boxed{+/-}$ key to your Calculator Tool Kit.

You will fill in the other boxes as you need to use those keys in the next few chapters.

OP-30.   **Definitions: base, exponent, and exponential form**
In the expression 5^3, **5** is called the **base**, and **3** is called the **exponent**.
We say 5^3 is written in **exponential form**. The expression 5^3 means $5 \cdot 5 \cdot 5$, and the 5's are called **factors**. The value of 5^3 in standard form is 125.

a)   Copy the definitions for base, exponent, and exponential form in your Algebra Tool Kit.

b)   Write 64 in exponential form, and then identify the base and the exponent.

OP-31. **Positive and Negative Powers of 2** You know how to compute powers of 2 for positive integer exponents. For example, 2^6 means $2 \cdot 2 \cdot 2 \cdot 2 \cdot 2 \cdot 2$, so by using multiplication you get $2^6 = 64$. The values of some powers of 2 have been calculated and entered in this table:

Power of 2	Numeric Value	Fraction Form	Expanded Form
2^6	64		$2 \cdot 2 \cdot 2 \cdot 2 \cdot 2 \cdot 2$
2^5	32		$2 \cdot 2 \cdot 2 \cdot 2 \cdot 2$
2^4	16		$2 \cdot 2 \cdot 2 \cdot 2$
2^3	8		
2^2			
2^1			
2^0			
2^{-1}		$\frac{1}{2}$	
2^{-2}			$\frac{1}{2 \cdot 2}$
2^{-3}	0.125		
2^{-4}			
2^{-5}			

In this problem, you will use patterning to complete the table and develop the meaning of negative integer exponents.

a)   First, look at the first four entries in the Numeric Value column: starting with 64, what calculation(s) can you make to get 32? Starting with 32, how can you get 16? Starting with 16, how can you get 8? Look for a pattern for how you can generate the rest of the column. Check your pattern for the entry that follows 8: does it give the same result as 2^2?

b)   On your copy of the OP-31 Resource Page, use the pattern you found in part (a) to generate the remaining entries in the table.

c)   Use your completed table to predict the value for 2^{-7}. What is the expanded form for 2^{-7}? What is the decimal value of 2^{-10}? What is the expanded form for 2^{-10}? Explain how you got your answers.

OP-32. Use the class number to write five more Five-Digit expressions that will generate the numbers from 21 through 25.

OP-33.   Record the amount you spent on books this semester to the nearest $1. If you have not finished purchasing your textbooks, predict the amount you will spend. You may need to check some prices in the bookstore or with your classmates. Also record the number of course units in which you are enrolled. This is important because you will need this information at the next class meeting.

OP-34.   Copy and complete each of these Diamond Problems:

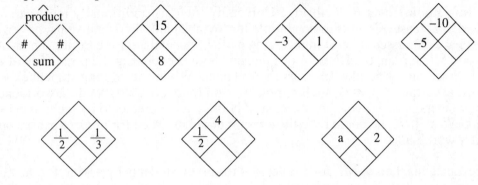

OP-35.   Use a Guess and Check Table to solve the following problem. Then state your solution in a sentence.

Find two consecutive even numbers whose sum is 142.

1.4 PREDICTING BOOK COSTS: DATA AND GRAPHS

OP-36.   On the coordinate grid posted on the wall, above the number of course units in which you are enrolled, place a sticky dot to indicate the amount you spent on textbooks this semester.

$$■$$▥$$■$$▥$$■$$▥$$■$$▥$$■$$▥$$■$$▥$$■$$

OP-37.   **Predicting Book Costs** At Oxbridge University students traditionally must carry a workload of at least 12 units to be eligible for financial aid. In recent years, however, students have expressed interest in receiving partial financial aid for carrying a part-time workload. Mr. Jordan, the Director of Financial Aid, is exploring the possibility of awarding stipends based on the number of course units taken. While considering student expenses, he reasons that certain expenses, such as housing and food, do not depend on workload. Mr. Jordan wonders, though, whether the cost of books can be predicted by the number of units a students takes. If so, he could base the amount of aid awarded for book expenses on each student's workload.

To investigate his theory, Mr. Jordan decides to survey students by asking for the number of course units in which they're enrolled and the amount each one spent on books to the nearest dollar. The result is the class graph generated in OP-36.

a)   After looking at the Predicting Book Costs graph, Mr. Jordan thinks a table that represents the data would be helpful. Discuss with your group how you could make effective use of your team so that each member can make an accurate table for Mr. Jordan. Make a table that represents the data from the Predicting Book Costs graph on the wall. You should make your own data table and verify the entries with your group before writing them down.

b)   Carry out your group's plan from part (a) so that each group member has a table of Predicting Book Costs data. Verify the entries in your table with group members.

OP-38.   Use the table you made for Predicting Book Costs to make your own graph of the class data on graph paper. On the horizontal (Number of Course Units) axis, use a scale where one tick mark represents 1 unit of workload. On the vertical (Cost of Books) axis, let each tick mark represent $5 in costs. It is important that each scale is consistent: tick marks on an axis should be the same distance apart, and should represent the same amount.

OP-39.   Use your Predicting Book Costs graph to answer the following questions:

a)   What is the range in book costs; that is, what are the smallest and largest book costs represented? What are the least and greatest number of course units represented?

b)   Could you use the graph to predict book expenses for someone who is taking 5.5 units? If so, explain how. If not, explain why not.

c)   What is the average number of course units being taken by students in the class? Explain and show how you found the average.

d)   What is the average amount of money spent on books by students in the class? Explain and show how you found the average.

OP-40.   Based on your Predicting Book Costs graph, was the Financial Aid Director's theory correct? Can you predict expenses for books if you know the number of course units a student is taking? Explain your response in complete sentences.

OP-41.   **$$$ More Dollar Bill Facts** Complete the following table using these facts: 454 US. one-dollar bills weigh about one pound, a stack of 454 one-dollar bills is about 4.96 centimeters thick, there are 100 centimeters in a meter, and there are 1,000 meters in a kilometer.

The <u>weight</u> of a one-dollar bill in pounds is _____

The <u>length</u> of a one-dollar bill in centimeters is   15.6 cm
The <u>length</u> of a one-dollar bill in meters is   _____
The <u>length</u> of a one-dollar bill in kilometers is   _____

The <u>width</u> of a one-dollar bill in centimeters is   6.63 cm
The <u>width</u> of a one-dollar bill in meters is   _____
The <u>width</u> of a one-dollar bill in kilometers is   _____

The <u>thickness</u> of a one-dollar bill in centimeters is _____
The <u>thickness</u> of a one-dollar bill in meters is   _____
The <u>thickness</u> of a one-dollar bill in kilometers is _____

OP-42.   Use a Guess and Check Table to solve the following problem. State your solution in a sentence.

Find four consecutive integers such that the product of the two largest is 46 more than the product of the two smallest integers.

OP-43.   Look at your table of powers of 2 from OP-31. How is the expanded form for 2^{-4} like the expanded form for 2^4? How are the expanded forms for 2^{-4} and 2^4 different?

OP-44.   **Powers of 10**
a)   Copy and complete the following table.

Negative Power	Decimal Form	Fraction Form	Positive Power
10^{-1}	0.1	$\frac{1}{10}$	$\frac{1}{10^1}$
10^{-2}	0.01		
10^{-3}		$\frac{1}{1000}$	
10^{-4}			

b)   What is the decimal form for 10^{-5}? What is the fraction form for 10^{-5}? What is the decimal form for 10^{-7}? What is the fraction form for 10^{-7}? Explain how you got your answers.

OP-45.   Imagine that you have an inquisitive cousin in New York who is eager to understand negative integer exponents. Based on your work so far, how would you explain negative integer exponents to your cousin?

OP-46.   Decide whether each equation is true or false. If false, write the correct result.

   a) $4 + 5 \cdot 6 = 54$

   b) $2 - (-3) + -17 = -12$

   c) $6 \cdot 6 - 6 \cdot 6 = 0$

   d) $4 + 2 \div (-1) = -6$

   e) $4 + -2(3) \div -3 = 6$

OP-47.   Use mental math (no paper and pencil allowed) to solve each of the following equations.

   a) $x + 3 = 7$ e) $x + 3 = -1$

   b) $3x = -12$ f) $-24 = -3x$

   c) $x - 3 = -7$ g) $10 = x + 14$

   d) $8x = 4$ h) $0 = 5x$

1.5 THE NATIONAL DEBT: CALCULATORS AND SCIENTIFIC NOTATION

OP-48.   **$$$$ The National Debt** Sara read that in 1993, the US. population was about 251 million, and the national debt was about $4.35 trillion. She tried to use a four function calculator to figure out what one citizen's share of the national debt would be, but it was no help. She also got stuck when she tried to enter 4.35 trillion in a scientific calculator in standard mode.

a)   Try to enter 4,350,000,000,000 in a scientific calculator in standard mode. Describe what happens.

b)   A convenient way to express very large (or very small) numbers is to use **scientific notation**. For example, 1993 written in scientific notation is $1.993 \cdot 10^3$ and 251,000,000 expressed in scientific notation is $2.51 \cdot 10^8$.

To use scientific notation on a scientific calculator, first put it in **scientific mode**: Then try one of the following sets of instructions:
- for a TI, either   press the $\boxed{\text{2nd}}$ key and then the $\boxed{\text{SCI}}$ key, or press the $\boxed{\text{3rd}}$ key and then the $\boxed{\text{SCI}}$ key; or
- for a Casio, press the $\boxed{\text{MODE}}$ key, then $\boxed{8}$, and then $\boxed{8}$ again; or
- check your calculator manual.

Now enter 1993 and press $\boxed{=}$. What does the calculator display? Explain the shorthand notation your scientific calculator uses for scientific notation.

c)   Large numbers such as "4,350,000,000,000" are troublesome because they do not fit in a scientific calculator display. Here's how scientific notation can help:

First express 4.35 trillion in scientific notation: $4.35 \cdot 10^{12}$.

Then enter it in the calculator by entering 4.35 and then $\boxed{\times}$, $\boxed{1}$, $\boxed{0}$, $\boxed{y^x}$, $\boxed{1}$, $\boxed{2}$, and $\boxed{=}$.

If in 1993 everyone in the US. gave the US. government $19,930, how close would the total contribution be to paying off the National Debt?

d)   In your Calculator Tool Kit, write how to enter a number in scientific notation in your calculator.

OP-49.   **Heavy Debt**

a)   If in 1993 all the people in the US. had to contribute equal amounts to pay off US. national debt, about how much would each person's share have been? Use what you know about order of operations and the $\boxed{(}$ $\boxed{)}$ keys on your calculator to make sure your answer is not bigger than the national debt.

b)   While she was thinking about the 1993 US. national debt, Sara recalled that 454 US. one-dollar bills weigh about one pound. If you paid your share of the 1993 national debt in one-dollar bills, about how much would it weigh? Could you carry one share?

OP-50.   **$$ Big Bucks** Your rich uncle has just died and has left you $1 billion. If you accept the money, you must count it for eight hours a day at the rate of $1 per second. When you are finished counting, the $1 billion is yours, and then you may start to spend it.

a)   Should you accept your uncle's offer? Why or why not?

b)   How long will it take to count the money?

OP-51.   a)   If you spent $1 million at the rate of $1,000 per day, how long would it take you to spend it?

b)   How long would it take to spend $1 billion at the same rate?

OP-52.   If 454 one-dollar bills weigh about one pound, …

	Standard Form	Scientific Notation
a) … about how much would 1,000,000,000 one-dollar bills weigh in pounds?		
b) … about how much would 1,000,000,000 one-dollar bills weigh in ounces?		

OP-53.   Use the data from OP-41 to compute about how far (in kilometers) one billion one-dollar bills would extend if they were …

	Standard Form	Scientific Notation
a) … strung side by side length-wise.		
b) … strung side by side width-wise.		
c) … stacked one on top of each other.		

OP-54.   In OP-31 you looked at patterns in expanded forms of powers of 2 to develop a meaning for negative integer exponents. You can use the $\boxed{x^y}$ or $\boxed{y^x}$ key on a scientific calculator to check your work. (If you need help, look in your Calculator Tool Kit.)

For example, to check that $2^{-1} = \frac{1}{2}$, you could use the $\boxed{y^x}$ key to get $2^{-1} = 0.5$. To find the decimal value for $\frac{1}{2}$, you could use division: $\boxed{1}$, $\boxed{\div}$, $\boxed{2}$, $\boxed{=}$.

Power of 2	2^6	2^5	2^4	2^3	2^2	2^1	2^0	2^{-1}	2^{-2}	2^{-3}	2^{-4}	2^{-5}
Decimal Form	64	32	16	8	4	2	1	0.5				
Fraction Form								$\frac{1}{2}$	$\frac{1}{4}$	$\frac{1}{8}$	$\frac{1}{16}$	$\frac{1}{32}$

a)   Use the $\boxed{y^x}$ key to find a decimal form for $2^{-2}, 2^{-3}, 2^{-4}$, and 2^{-5}.

[PROBLEM CONTINUED ON NEXT PAGE]

OP-54. continued

b)   Describe how to use a simple four function calculator to find a decimal form for $\frac{1}{4}$.

c)   Find the $\boxed{1/_x}$ key on your scientific calculator and explore how you could use it to find a decimal form for $\frac{1}{4}$. Try it with $\frac{1}{8}$, $\frac{1}{16}$, and $\frac{1}{32}$. Write a sentence to describe what you discover about using the $\boxed{1/_x}$ key.

OP-55.   What is the highest power of 2 your calculator will compute? What is the lowest power of 2 your calculator will compute? Be prepared to explain how you figured these out.

OP-56.   Here are the first five figures in a geometric pattern:

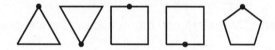

a)   Copy the five figures shown on your paper.

b)   Draw four figures that would follow the first five figures you copied.

c)   How many sides will the 16th figure have?

d)   Where will the dot in the 27th figure appear?

e)   Explain, so that a new student in the class would understand, how you determined your answers for parts (c) and (d).

OP-57.   **Progress Reports** Jack, Donna, Nancy, and Lee are students in Ms. Speedi's math class. Each student's progress report is represented by one of the points on the graph below.

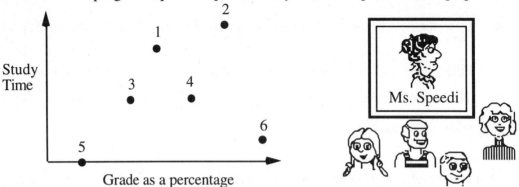

"Jack never studies and has a poor test percentage."

"Nancy is very able, but because of her active social life, she seldom studies outside of class. Her test percentage is OK, but could be better."

"Donna has worked a lot both in and out of class. Her test grades are very good."

"Lee studies often outside of class and has done reasonably well on the tests."

a)   Which report corresponds to which point?

b)   Make up reports for the remaining two points on the graph.

c)   Think about your own study skills and predict your math grade at the end of the first quarter. Copy the graph and place a point on it that reflects your prediction and your study habits. Explain your choice of location.

OP-58.   Use a Guess and Check Table to solve the following problem. State your solution in a sentence.

The length of a rectangle is 4 meters more than twice its width. If the area of the rectangle is 126 square meters, find its length and width.

OP-59.   Make each of the following calculations using mental math. No paper, pencil, or calculator allowed! Write down the results of your calculations in complete statements: $17 - (-12) = 29$.

a) $32 + (-7)$ b) $-5 + (-10)$

c) $[-13 + (-12)](-4)$ d) $(-8)(-8) - (-8)$

e) $-1\frac{1}{2} + \frac{1}{2}$ f) $\frac{1}{2}(-68)$

g) $1\frac{1}{2} + \frac{3}{4}$ h) $\frac{3}{4}(100)$

OP-60.   Which, if any, of the calculations in problem OP-59 were challenging to do without a calculator? Explain why.

OP-61.   You've been working in groups for several class meetings now. As you get to know each other, you will probably develop styles of working together that are unique to your group. What works for one group may not for another. However, it may help you develop as a group to consider methods that have worked for others.

Some groups collaborate effectively when each of the following roles is assumed by someone in the group, although no single person should occupy a particular role for more than a few minutes. As you read the description of each role, think about the way your group has been working together to solve problems.

Questioner: "What are we supposed to do?" "I don't understand." "Where do I find the paper clips we need?" "What does _____ mean?"

Organizer: "I'll get the paper and colored pencils; you get the rulers." "You do the first two parts of the problem, and I'll do the next two parts." "All right, let's get this chart made."

Prober: "Hey, wait a minute! Maybe there's a pattern in this table." "What if we put it in this order?" "Maybe this works for all of them." "Can we write an equation?"

Summarizer: "Let's look at what we've done." "Remember to label the axes." "Do you have all three pages together, Jennifer?" "Did we answer all the questions?" "Don't forget--we need to write our solution in a sentence."

Choose a problem that your group has worked on in class and recall how your group worked together. Identify the problem you've selected, and then use complete sentences to describe what happened in your group while you were solving the problem:

a)   Which of the roles described above did you take on? Which was the most comfortable for you, and why? How about the other members of your group?

b)   In your group, what worked well? What could you do to help the group work more effectively?

1.6 DRIVING TO REDWOOD CITY: DATA FROM A GRAPH

OP-62.   Ms. Speedi drove from Ukiah to Redwood City by the route shown on the map below. The graph depicts her trip.

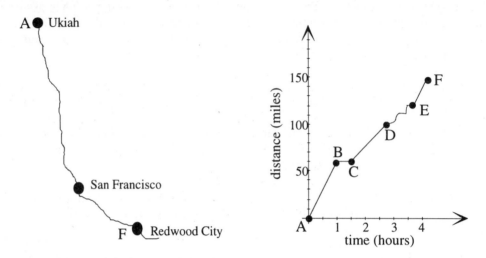

a)   How far did Ms. Speedi travel on her trip?

b)   Describe the car's speed during the interval from A to B.

c)   What might have happened during the interval from B to C ?

d)   Something different happened as Ms. Speedi traveled from D to E. Explain what you think might have happened.

e)   In going from C to D, did Ms. Speedi travel faster or more slowly than she did from A to B ? Explain how you know.

OP-63.   Imagine you are driving south on Interstate 5.

a)   If you drive at an average speed of 60 miles per hour for two hours, how far will you go?

b)   If you drive at 65 mph for three hours, how far will you travel?

c)   If you drive at 48 mph for 30 minutes, you will be on the road for only one-half of an hour. How far will you travel?

d)   If you drive at 57 mph for 40 minutes, for what fraction of an hour will you travel? How many miles will you travel?

e)   Explain in words the relationship between your car's speed, the amount of time on the road, and the number of miles traveled.

OP-64.   Whenever the California Highway Patrol (CHP) follows Tony Ticket he drives 64 miles per hour (mph). When the CHP isn't around, though, he travels at 80 mph. Recently, Tony drove 456 miles to Los Angeles. During this trip the CHP followed Tony for $1\frac{1}{2}$ hours more than the time he drove at 80 mph.

   a)   Which is more, the amount of time Tony drove at 64 mph, or the amount of time he drove at 80 mph?

   b)   If Tony drove at 64 mph for 3 hours and at 80 mph for 1.5 hours, how far did he drive in all?

   c)   Use a Guess and Check Table to find how many hours Tony Ticket drove at 80 miles per hour. State your answer in a sentence.

OP-65.   a)   Create two more Diamond Problems of your own and solve them on your homework paper. For example:

   b)   Can you create a Diamond Problem for which all four numbers would be negative? If so, do it. If not, explain why not.

OP-66.   Compute:

   a) $-50 - 30$

   b) $4(-9 - 6)$

   c) $(-13)(-2)$

   d) $-178 - (-3)$

   e) $-\frac{1}{2} - 2\frac{3}{4}$

   f) $(-2)^2 + \frac{2}{3}(-\frac{1}{2} + (-\frac{1}{6}))$

   g) $(2.35)(-4.01)^2$

   h) $(0.005) \div (-0.021) + 8.31$

OP-67.   ⇌ **Car Comparison** The following graph describes two cars, Car 1 and Car 2.

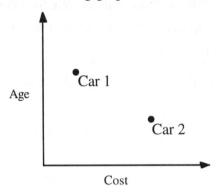

a)   Which car is the most expensive?

b)   Which car is cheaper?

c)   Which car is older?

d)   Name two cars (make and year) that you think would fit the graph.

OP-68.   ⇌ ⇌ **More Car Comparison** The following three graphs describe two cars, Car A and Car B. ("Range" means the distance a car can travel on one tank of gas.)

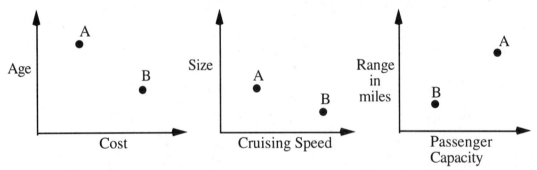

Decide whether each of the following statements is true or false. Explain your reasoning.

a)   The newer car is more expensive.

b)   The slower car is larger.

c)   The larger car is newer.

d)   The less expensive car carries more passengers.

OP-69.   Use a Guess and Check Table to solve the following problem. State your answer in a sentence.

The sum of three numbers is 61. The second number is five times the first and the third is two less than the first. Find the three numbers.

OP-70. **Chapter 1 Summary: Rough Draft** In this course, as in most courses you study, you will continue to work on many ideas, concepts and skills. There is a lot of information in this course. It may be difficult to process this information and organize it for yourself, so that you can easily remember **all** of it. To help remember a lot of information, it helps to summarize it in the context of big ideas.

You will be asked to write a summary for every chapter in this course. This is an important process in learning. Therefore, this problem, along with OP-71 (group discussion) and OP-85 (revision), is a very important one and your final presentation should be thorough and neat. For now, though, you need to brainstorm ideas, write them down, and be ready to discuss them with your group at the next class meeting.

It may be that you will be able to use your summary to help on quizzes. Check with your instructor about this.

The main ideas of Chapter 1 are:

- working collaboratively to solve problems;

- collecting and organizing data or information;

- using a Guess and Check Table to solve word problems; and

- reading and interpreting graphs.

You also worked with problems that involve the following topics:

- computing with integers;

- rewriting expressions using the meaning of integer exponents;

- using your calculator more efficiently; and

in the next section you will be

- calculating simple probabilities.

On separate sheets of paper: There are a lot of ideas covered in this chapter. Some ideas should be review while others may be new to you. Answer the following questions completely and be ready to discuss them with your group at the next class meeting. You will have the opportunity to revise your work, so at this point you should focus on the content of your summary rather than the appearance.

a) For each of the four main ideas of the chapter, find at least one problem which illustrates the idea. Be sure to include a description of the original problem and a completely worked out solution. Also write one or two sentences describing each idea.

b) Of the other four ideas, choose one which you understand better now than before you worked through this chapter, and describe what you learned.

c) Write a realistic self-evaluation related to the second set of skills in the list above. Which ones do you feel confident in performing? Which skills do you still need to work on? Decide if any of the ideas listed need to be included on your Tool Kit.

1.7 SUMMARY AND REVIEW

OP-71.   **Chapter 1 Summary: Group Discussion** Take out your draft of the Chapter 1
Summary from OP-70.

a)   For each of the four main ideas of the chapter, choose one member of the group to lead a
short discussion. The discussion leaders should take turns

• explaining the problem they chose to illustrate their main idea, and

• explain why they chose that particular problem.

This is a good opportunity for you to make sure your chapter summary is complete --
you will revise it for homework.

b)   Discuss the following questions with your group: With which of the eight topics listed in
OP-70 do you feel most confident in your understanding? Which topics do you still need
to work on? (You will have more opportunities for practicing **all** of the skills from
Chapter 1 in the upcoming chapters.)

c)   Make sure your Tool Kit is updated and correct. Add any important ideas that arose in
your group's discussion. Check the accuracy of your Tool Kit entries.

OP-72.   Suppose you toss three coins, a penny, a nickel, and a dime. One possible outcome is for the
coins to come up "heads, heads, heads" (H, H, H).

a)   What other possible outcomes might occur? List **all** the possibilities you can find,
including H,H,H.

b)   How can you be <u>sure</u> that you listed all the possibilities in part (a)? Describe a strategy
for checking your list.

c)   Copy the following definition in your Algebra Tool Kit:

Definition: When all outcomes are equally likely to occur, the **probability** (or likelihood)
that a particular event occurs, denoted by P(event), is the fraction

$$P(\text{event}) = \frac{\text{number of ways that the event occurs}}{\text{total number of possible outcomes}} .$$

For example, when rolling a fair standard die, the probability of getting an odd number is
written $P(\text{"odd"}) = \frac{3}{6}$.

d)   If you randomly toss three coins, what is the probability that exactly one coin comes up
"heads?"

e)   If you randomly toss three coins, what is the probability that at least one coin comes up
"heads?"

OP-73. a) Copy these three figures and then draw five figures that could reasonably follow in the sequence below. State in a sentence or two the rule for your pattern.

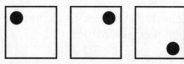

Figure 1   Figure 2   Figure 3

b) Without drawing any more figures, determine how Figure 27 should be drawn. Explain how you made your decision so a student new to the class could understand.

OP-74. a) Copy each pair of axes. On each graph label two points that would represent Cars A and B as described in the Car Comparison Problem.

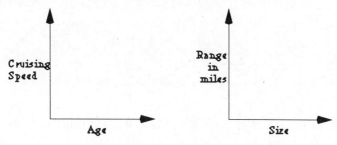

b) Which car would you buy? Why? Explain your answer a complete sentence.

OP-75. Express three numbers between 30 and 40 using the class number in the Five-Digit manner as done in problem OP-10.

OP-76. If the class number had been 6, on what lines would each expression go?

a) $6 \cdot 6 + \dfrac{6}{6} - 6$ b) $\dfrac{6 + (6 - 6) \cdot 6}{6}$ c) $6 + 6^{[(6+6)\div 6]}$

OP-77. Copy the axes for problem OP-57, and put a dot to indicate each of the following students.

Student A, who studies hard but gets only average grades.

Student B, who studies little but gets good grades.

OP-78. Compute:

a) $-15 + 7$ b) $-50 - 30$ c) $-50(-30)$

d) $(-2) \div (-25)$ e) $\dfrac{3}{8}\left(-46 - 27\left(\dfrac{-2}{9}\right)\right)$ f) $(-6 + 17) - 20$

g) $[-12 + (-18)] - 15$ h) $(-2)(-2)(-2)(-2) - 2$ i) $(4982)(-556)(0)$

OP-79. a) Which of the calculations in problem OP-78 could be done without the aid of a calculator?

b) Which of the calculations in problem OP-78 would be difficult or time-consuming to do without a calculator?

OP-80.   Use a Guess and Check Table to solve the following problem. State your solution in a sentence.

A cable 84 meters long is cut into two pieces so that one piece is 18 meters longer than the other. Find the length of each piece of cable.

OP-81.   Copy and solve these Diamond Problems:

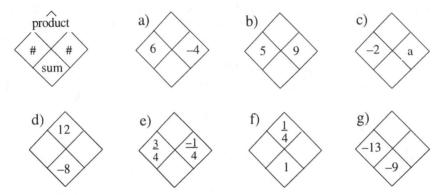

OP-82.   Use a Guess and Check Table to solve the following problem. State your solution in a sentence.

The total cost for a chair, a desk, and a lamp is $562. The desk costs four times as much as the lamp and the chair costs $23 less than the desk. Find the cost of the chair and the desk.

OP-83.   Ms. Speedi carries only bills in her book bag, no coins. As she raced off to work this morning, she tossed in some ones, fives, and tens. There are eight bills in all and their total value is 38 dollars.

a)   How many of each kind of bill are in Ms. Speedi's bag?

b)   If Ms. Speedi randomly grabs one bill out of her book bag, what is the probability that it is a five dollar bill? Explain how you know.

OP-84.   Throughout this chapter you worked in your group daily. Write two paragraphs to address the following questions:
        What role did you play in the group?
            Were you a leader, taking charge?
            Did you keep your group on task?
            Did you ask questions?
            Did you just listen and copy down answers?
            In what ways are you a good group member?
            How could you do better?
        How well did your group work? Explain.

OP-85.   **Chapter 1 Summary: Revision** This is the final summary problem for Chapter 1. Using your rough draft from OP-70 and the ideas you discussed in your groups from OP-71, spend time revising and refining your Chapter 1 Summary. Your presentation should be thorough and organized, and should be done on a separate piece of paper.

1.8 DEBITS AND CREDITS: INTRODUCTION TO INTEGER TILES (OPTIONAL)

OP-86.   **Representing Numbers with Integer Tiles**   You've worked with integers before and have probably memorized some "rules" for computing with positive and negative values. Although you may already be able to get the answers, your understanding of the symbolic operations can be reinforced by using pictures or models.

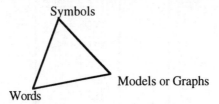

Integer tiles can be used to represent, or *model*, addition, subtraction, and multiplication with integers. Using the tiles may help you increase your understanding of integer operations.

Integer tiles have two distinct sides (or "**+**" and "**–**" shapes) to represent "positive" and "negative." You'll use them in a simple model of an accounting balance sheet where a "**+**" tile is used to record a credit of $1 and a "**–**" tile is used to record a debit of $1.

The key idea in this model is that a positive tile and a negative tile "nullify" each other; in other words, a "**+**" tile combines with a "**–**" tile to make zero. By accounting for these "zeros," you can represent one "balance" in many different ways.

Before you can use the tiles to show calculations with integers, you need to know how to represent a single integer. Here are three different ways to represent a net balance of $2:

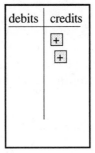   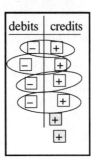   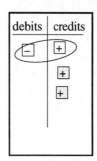

$2 net balance $2 net balance $2 net balance

Here are three different ways to represent a net balance of –$1:

   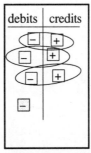   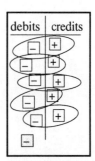

–$1 net balance   –$1 net balance   –$1 net balance

[PROBLEM CONTINUED ON NEXT PAGE]

OP-86. continued

It is important to be able to represent a **zero balance** in a number of different ways (the first
step in modeling a calculation with integer tiles is to start with a zero balance). To model a net
balance of $0, use *equal numbers* of positive and negative tiles:

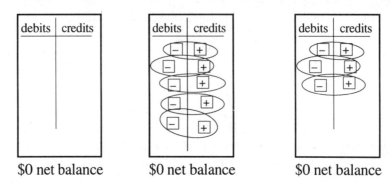

$0 net balance $0 net balance $0 net balance

To record your work, make a simple sketch using "**+**" for a positive tile and "**–**" for a negative
tile. To record the balance sheet on the left, you could make the drawing on the right.

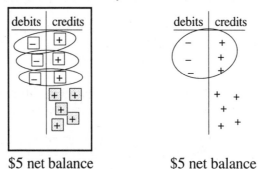

$5 net balance $5 net balance

Use integer tiles and an Integer Tiles Balance Sheet (see Resource Pages) to represent each of
the following integers in three different ways. After building each model with tiles, record
your work by sketching the balance sheet using "**+**"'s and "**–**"'s.

a)   5   b)   2   c)   –7   d)   0

OP-87.   When representing calculations, it sometimes will be useful to start with a zero balance made of
           a large number of tiles. Use at least 12 tiles to represent a balance of 5.

OP-88.   **Addition with Integer Tiles** You can use integer tiles and a debit/credit balance sheet to model addition with integers. For example, here is how to model 4 + –3:

Start with a zero balance (Step 1 below). Then add the appropriate number of tiles -- in this example, add 4 "**+**" tiles, and then 3 "**–**" tiles -- to the zero balance (Steps 2 and 3). Next, match pairs of positive and negative tiles (push them together) to account for "zeros" (Step 4). Finally, record your results as in Step 5.

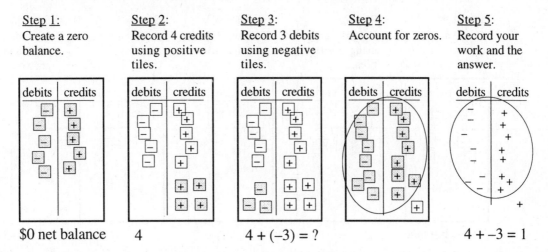

Model each sum using integer tiles. Sketch each model and write the problem with its answer below your drawing.

a)   –7 + (–2)   c)   3 + (–4)   e)   2 + 5

b)   –1 + 5   d)   –4 + 0   f)   7 + (–8)

OP-89.   To model subtraction of integers with tiles, you can physically remove the appropriate number of positive or negative tiles from a debit/credit balance sheet. The tricky part is making sure you've got enough tiles to remove.

For example, suppose you want to model 3 – 7 and you start with no tiles on your debit/credit balance sheet. After recording 3 credits, you would want to *remove* 7 credits. *BUT there aren't enough "+" tiles on the balance sheet for you to remove seven!* So start again …

At the start of the problem, in what way could you show a zero balance that would allow you to remove 7 credits from a balance of 3 credits? With your group, find a zero balance that works, and then draw and describe the corresponding balance sheet.

OP-90.   **Subtraction with Integer Tiles**   Here's how to use integer tiles and a debit/credit balance
sheet to model subtraction with integers. For example, for 2 – (–3):
Start with a $0 net balance (Step 1 below). Represent the first number -- in this example the
"2" -- with tiles (Step 2). To subtract, *physically remove* the appropriate number of positive or
negative tiles. In this example, we would remove three negative tiles (Step 3). Finally, match
pairs of positive and negative tiles (push them together) to account for zeros (Step 4). Record
your work as in Step 5.

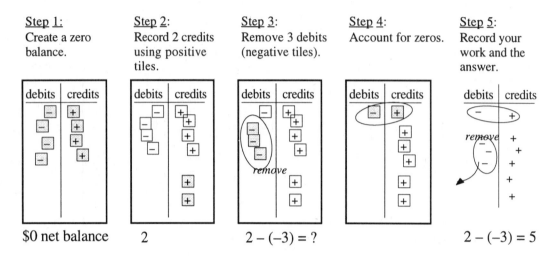

Model each difference (subtraction problem) using integer tiles. Sketch each model and write
the problem with its answer below your drawing.

a)   3 – 5   c)   5 – 3   e)   –2 – 1

b)   –4 – (–6)   d)   4 – (–4)   f)   –4 – 4

OP-91.   **Multiplication with Integer Tiles**   To model multiplication using integer tiles on a
debit/credit balance sheet, you can use the notion that multiplication can be represented by
repeated addition. For example, 2(3) = 3 + 3 and 2(–3) = (–3) + (–3). In terms of the
integer tiles,

$\quad$ 2(3)   means add 2 groups of 3 positive tiles;
$\quad$ 2(–3)   means add 2 groups of 3 negative tiles;
$\quad$ –2(3)   means *remove* 2 groups of 3 positive tiles; and
$\quad$ –2(–3) means *remove* 2 groups of 3 negative tiles;

When using the tiles, first create a zero balance **using enough tiles** to allow you to carry out
the multiplication. You may not know how many "enough" is at this point but that's okay--
you'll find out in the second step when you try to add or remove **groups** of tiles. If needed,
you can go back and add more tiles to the original zero balance.

[PROBLEM CONTINUED ON NEXT PAGE]

OP-91. continued

Example 1. To model 2(–3), first create a zero balance (as in Step 1 below) using a sufficient number of tiles. Then add 2 groups of 3 negative tiles (Step 2). Finally, account for "zeros" as usual (Step 3), and then record your work as in Step 4.

Step 1:
Create a zero balance.

Step 2:
Record 2 groups of three debits (negative tiles).

Step 3:
Account for zeros.

Step 4:
Record your work and the answer.

$0 net balance   2(–3) = ?   2(–3) = –6

Example 2. To model –2(3), start with a zero balance (as in Step 1 below) made with a sufficient number of tiles. Now remove 2 groups of 3 positive tiles (Step 2). Finally, account for "zeros" as usual (Step 3), and then record your work as in Step 4.

Step 1:
Create a zero balance.

Step 2:
Remove 2 groups of three *credits*.

Step 3:
Account for zeros.

Step 4:
Record your work and the answer.

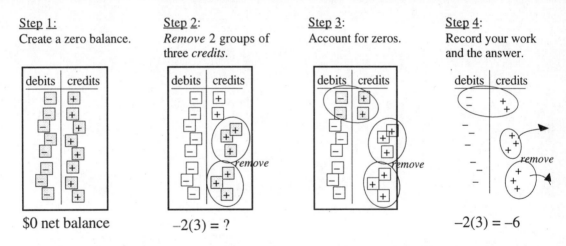

$0 net balance   –2(3) = ?   –2(3) = –6

a)   Look again at each example. In Step 2, how were the tiles which formed the initial zero balance involved for each situation? When creating a zero balance to begin a calculation, how can you tell what a "sufficient" number of tiles is?

Model each product using integer tiles. Sketch each model and write the problem with its answer below your drawing.

b)   3(–5)   d)   5(–2)   f)   –2(5)

c)   –4(3)   e)   4(2)   g)   –4(–2)

OP-92.   a)   Copy and complete the balance sheet below to show −3 + (−1).

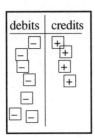

          b)   Copy and complete the balance sheet below to show −3 − (−4).

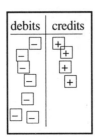

OP-10. The Five-Digit Problem

Generate the integers from 0 to 20 using any one digit five times along with any "legal"
mathematical operations (for example: +, −, ×, ÷, …).

0
1
2
3
4
5
6
7
8
9

OP-10. The Five-Digit Problem continued

10

11

12

13

14

15

16

17

18

19

20

Algebra Tool Kit: the "what to do when you don't remember what to do" kit

Order of
Operations

Integer
Arithmetic

base,
exponent,
exponential
form

negative
exponents

scientific
notation

Algebra Tool Kit: the "what to do when you don't remember what to do" kit

probability

perimeter

area

discrete
vs.
continuous

Centimeter Grid Graph Paper

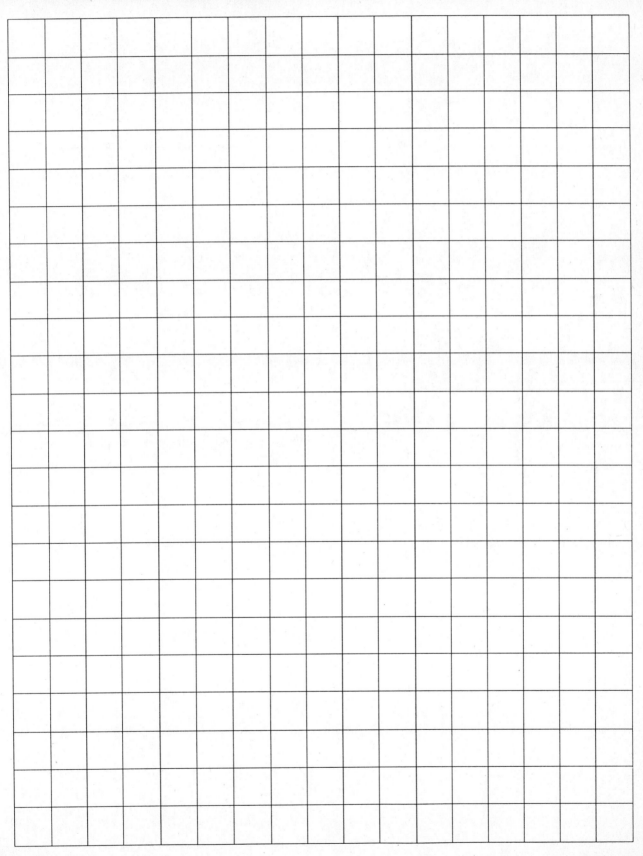

Calculator Tool Kit: for instructions on using your calculator

$\boxed{+/_-}$	Changing the sign of a number	

$\boxed{y^x}$	Raising a number to a power	

$\boxed{EE}$ or $\boxed{EXP}$	Entering a number in scientific notation	

$\boxed{a^b/_c}$	Working with fractions	

$\boxed{\frac{1}{x}}$	Finding the reciprocal of a number	

Calculator Tool Kit: for instructions on using your calculator

OP-31. Positive and Negative Powers of 2

Power of 2	Numeric Value	Fraction Form	Expanded Form
2^6	64		$2 \cdot 2 \cdot 2 \cdot 2 \cdot 2 \cdot 2$
2^5	32		$2 \cdot 2 \cdot 2 \cdot 2 \cdot 2$
2^4	16		$2 \cdot 2 \cdot 2 \cdot 2$
2^3	8		
2^2			
2^1			
2^0			
2^{-1}		$\dfrac{1}{2}$	
2^{-2}			$\dfrac{1}{2 \cdot 2}$
2^{-3}	0.125		
2^{-4}			
2^{-5}			

Integer Tiles

Glue the entire grid of integer tiles to a piece of lightweight cardboard (an empty cereal box works well) or another piece of paper. Carefully cut out the individual tiles and keep them in a re-sealable envelope or plastic bag. Store the bag of tiles in your notebook.

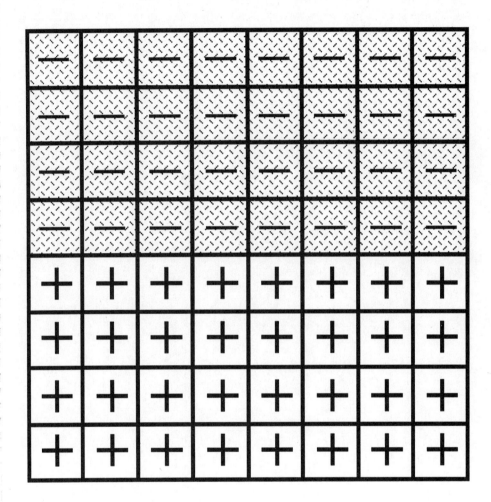

Integer Tiles Balance Sheet

Debits	Credits

Chapter 2

MS. SPEEDI:
Patterns
and Graphs

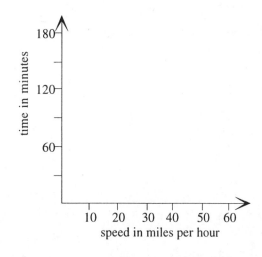

CHAPTER 2

MS. SPEEDI:
PATTERNS AND GRAPHS

Chapter 2 introduces patterning, an important problem solving strategy that you will use throughout this course. You will look for and describe patterns and use them to solve a variety of problems. As part of the problem solving process you will use patterns to make conjectures and see the importance of checking conjectures. You will also apply your patterning skills to making graphs of rules, and will see that some rules have graphs that are lines, while others result in graphs of non-linear curves.

In this chapter you will have the opportunity to:

- use the important problem solving strategy of looking for, describing and using patterns.

- make and test conjectures.

- see how graphs, tables and rules are related.

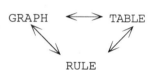

- graph different kinds of rules to produce lines, parabolas, and other curves.

CHAPTER CONTENTS

2.1 THE ALGEBRA WALK

SP-1.   **The Algebra Walk** Many of you have some experience with the xy-coordinate system from previous classes. Regardless of your background, however, the Algebra Walk is an unconventional approach that may help you add to your understanding of graphing by focusing on graphs from a new perspective.

The Algebra Walk is an exercise in "human graphing" in which people represent points on a graph. Your instructor will have marked off a large set of axes on the ground outside your classroom. Before going outside, you will receive at least one colored card marked with an integer. You will also need an Algebra Walk Observations sheet from your Resource Pages, a pencil, and a book or other hard surface on which to write while outdoors. During this exercise, you will assume two different roles--sometimes a grapher, and sometimes an observer. Graphers follow oral instructions on how to form their graph. Observers have three tasks: to write the graph's rule in words; to sketch what they see; and to describe in words what they see.

Instructions for Graphers: Look at the integer on your card. At a signal given by your instructor, find your place on the horizontal axis. Stand with both feet on your integer, facing in the positive "y" direction, and with your back to the rest of the class. Your instructor will call out instructions similar to the ones below. Follow the instructions to form the human graph.

> "Look at your number. Multiply it by 2. Add one. Got it? Write the resulting number on the right-hand side of your card.

> "When I say 'Go,' walk that number of spaces forward or backward, depending on whether your result is positive or negative. (A 'space' is the distance between two tick marks on the y-axis.) Ready? GO!"

Keep your card with both the starting integer and the resulting number; you will need them when you return to the classroom.

Instructions for Observers: Find a place where you can see the entire outdoor grid and face the positive "y" direction. Listen to the instructions to the graphers, and then on your Observations Sheet, write the graph's rule in words, **roughly** sketch the graph, and, as clearly as you can, describe in words the shape of the human graph.

Blue Rule:	Double your input number, then add one.
Green Rule:	Multiply your number by –2.
Yellow Rule:	Add four to your number.
Orange Rule:	Multiply your number by –1, and then add 4.
Red Rule:	Subtract one from the square of your number.

SP-2.   **Follow-Up to the Algebra Walk** When you return from the Algebra Walk, you will find several large grids on the walls of the classroom. Find the grid which matches the color of your integer card, and put a matching sticky dot to represent your "point" on the human graph.

Work in groups to complete the following graphs, using the Grid Sheet from your Resource Pages. Make each graph on a separate grid.

Note: Graphing can quickly become tedious because there are so many points to plot. People working alone can easily get so bogged in drudgery that they miss making important observations and connections. Here's an example where the use of "group power" can optimize a group's learning by minimizing tedium and redundancy.

a)   On a Grid Sheet, complete a table for the rule $y = 2x + 1$, then neatly graph each point.

IN (x)	−4	−3	−2	−1	0	1	2	3	4
OUT (y)									

b)   Now, repeat part (a) for each of the following rules, using a new grid for each one.

$y = -2x$

$y = x + 4$

$y = -x + 4$

$y = x^2 - 1$

c)   Compare the five graphs in parts (a) and (b). How are they similar? How are they different? Write several sentences to describe what you notice.

d)   Express each of the five symbolic rules in parts (a) and (b) in words.

SP-3.   Use your graphs from the Algebra Walk Follow-Up to answer the following questions.

a)   How can the graph for the rule $y = 2x + 1$ be used to predict the output value for an input (x-value) of 5 ? How can the graph be used to predict the output value associated with an input of $3\frac{1}{2}$?

b)   If you wanted an output of 7 for the rule $y = -x + 4$, what would you need as an input?

c)   For each of the rules in parts (a) and (b) of problem SP-2, where does the graph cross the y-axis? Add this information to your Grid Sheets.

SP-4.   a)   You know that x^3 means $x \cdot x \cdot x$. Explain what $(y^2)^3$ means.

b)   Rewrite the product $(y^2)^3$ as simply as possible without parentheses.

SP-5.   The graph below shows some information about the number of cans of soda in the soda machine at two times during an average day.

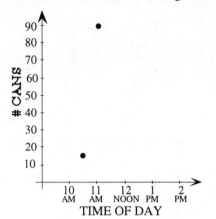

a)   What information is given by the two dots?

b)   What do you think happened between the times indicated by the two dots?

c)   If lunch is between noon and 12:45 pm, copy the graph and place a dot corresponding to the number of cans you think would be in the machine at 1:00 pm. Describe why you placed the dot where you did.

For problems SP-6 and SP-7:

a)   Make a table like the one below and complete the table using the rule provided. (You could also use a Grid Sheet from your Resource Pages.)

IN (x)	–4	–3	–2	–1	0	1	2	3	4
OUT (y)									

b)   Plot the points on a graph, then label the graph with its equation. Be sure to label the x-axis and y-axis, and to include the scale for each one.

SP-6. $y = -x + 2$

SP-7.   To find y, triple x and add 1.

SP-8.   Belinda made the table below, but then she lost the rule she used to calculate the y-values.

IN (x)	–4	–3	–2	–1	0	1	2	3	4
OUT (y)	17	10	5	2	1	2	5	10	17

a)   Plot the points Belinda found on a graph. Be sure to label the axes and to include the scale.

b)   Explain in words what Belinda did to the input value, x, to produce the output value, y.

c)   Use algebraic symbols to write the process you described in part (b).

SP-9.   Write a paragraph to describe how you think the Algebra Walk will help your understanding of graphing. Be as specific as you can.

SP-10.   Lawson shaded numbered squares to fit a certain pattern. The first five figures in his pattern
          are shown below.

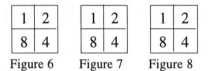

          Figure 1   Figure 2   Figure 3   Figure 4   Figure 5

          Study the figures and try to find Lawson's pattern.

          a)   Copy and shade the next three large squares to continue the pattern you found.

                    | 1 | 2 | | 1 | 2 | | 1 | 2 |
                    | 8 | 4 | | 8 | 4 | | 8 | 4 |
                     Figure 6   Figure 7   Figure 8

          b)   Describe your pattern in a sentence.

          c)   If you were to extend the pattern as far as you could, what would be the last figure in the
                sequence?

SP-11.   Use a Guess and Check Table to solve the following problem.

          The length of a rectangle is six more than the width. If the perimeter is 52, find the length and
          width of the rectangle. What is the area of the rectangle?

2.2 MS. SPEEDI'S COMMUTE

SP-12.   a)   Discuss and solve the following problem with your group:

On her way to a mathematics conference, Ms. Speedi figured that she was driving at an average speed of 58 miles per hour, and at that rate it would take her a total of three and half hours to get to the conference. How far did she drive?

b)   Describe what you did to solve part (a). Using that knowledge, work together to write a concise way of explaining the relationship between distance traveled, speed, and time.

SP-13.   ⏱ **Get me to work on time** Ms. Speedi, the math teacher, has figured out that if she drives to work at 60 miles per hour, it will take her half an hour to get to school. Unfortunately that speed exceeds the speed limit, so Ms. Speedi must drive at a slower rate. Some days she might drive at 30 miles per hour. Other days she could drive at 45 miles per hour. Ms. Speedi doesn't want to be late for work so she needs to know the speed and time combinations that might allow her to get to school on time.

a)   How many miles away from work does Ms. Speedi live?

b)   Calculate how long it would take Ms. Speedi to get to school if she drove at 30 miles per hour. Calculate how long it would take if she drove at 45 miles per hour. (Review your work on OP-69 and SP-12 if you need to.)

c)   Record your data from part (b) in a table. Add at least five more speeds to the data table and calculate the time associated with each speed. Then convert each time in hours to minutes.

speed	time (hours)	time (minutes)
60 mph	$\frac{1}{2}$ hr	
30 mph		
45 mph		

d)   Copy the axes, making them large enough to use up most of one side of a sheet of graph paper. Graph all of your data.

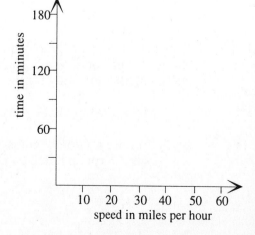

SP-14.   Use your graph of Ms. Speedi's data to answer the following questions. Remember to write complete sentences.

    a)   When reading the graph from left to right, does the graph slope always upward, always downward, or change directions?

    b)   Do you think the graph should cross the horizontal or vertical axis? If so, where? If not, why not?

    c)   Do the points lie in a straight line? Do you think the points should lie in a line? Explain your answer.

SP-15.   Suppose Ms. Speedi tells you that she has only 15 minutes to get to school. Describe how you could use your graph to quickly predict the speed at which she must drive. How could you find the speed more accurately? Write your responses in complete sentences.

SP-16.   If Ms. Speedi's gas pedal is stuck so that she can only drive 8 mph, how long will it take her to get to school? Use your graph of Ms. Speedi's data to make an estimate, and then calculate the time more accurately.

For problems SP-17 and SP-18:

    a)   Make a table like the one below (or use a Grid Sheet from your Resource Pages) and complete the table using the rule provided.

IN (x)	−4	−3	−2	−1	0	1	2	3	4
OUT (y)									

    b)   Plot the points on a graph, then label the graph with its equation. Be sure to label the x-axis and y-axis, and to include the scale for each one.

SP-17. $y = -x + 1$

SP-18. $y = x^2 - 2$

SP-19.   Lance is taking a geography quiz with only three questions. For each question, he must choose "True" or "False." Unfortunately, he missed the last two weeks of his geography class, and did not study for the quiz.

    a)   Make a list of all the possible answer sheets Lance could turn in.

    b)   If Lance randomly guesses answers, what is the probability that he earns 100% on the quiz? Explain how you know.

SP-20.   a)   Copy and complete the list below and look for patterns.

$$4^1 = 4$$
$$4^2 = 16$$
$$4^3 = \underline{\hphantom{00}}$$
$$4^4 = \underline{\hphantom{00}}$$
$$4^5 = \underline{\hphantom{00}}$$
$$4^6 = \underline{\hphantom{00}}$$

   b)   List the next three exponents for powers of four that will have a "6" in the ones place.

   c)   What number is in the ones place of the standard form for 4^{128} ? (Caution: Your calculator won't be of any help.)

   d)   Explain how you figured out the answer to part (c) so that a high school student could give the answer for 4^{997}.

SP-21.   Melissa cut a 150-centimeter board into two pieces. One piece is 24 centimeters longer than the other piece. How long is each piece of board? Use a Guess and Check Table to solve this problem.

SP-22.   Copy each equation, then find all values of n that make the equation true.
   a) $2^n = 16$ b) $n^2 = 64$ c) $3^n = \dfrac{1}{9}$

SP-23.   Write the rule $y = \dfrac{1}{2}x + 3$ in everyday words.

SP-24.   Copy and solve these Diamond Problems:

SP-25.   Use the meaning of integer exponents to write each of the following expressions as simply as possible <u>without exponents</u>. For example,

$$x^3 = x \cdot x \cdot x,$$
$$x(x^3) = x \cdot (x \cdot x \cdot x), \text{ and}$$
$$x^3 \div x = \frac{x \cdot x \cdot x}{x} = x \cdot x$$

   a) x^2 b) $x^2 \cdot x^5$

   c) $x^3(x^4)$ d) $x^3 \div x^2$

SP-26.   Use the meaning of integer exponents to re-write each of the following expressions with a
          single exponent. For example,

$$4 \cdot 4^2 = 4 \cdot (4 \cdot 4) = 4^3.$$

a) $(5^2)^3$ b) $(10^3) \cdot (10^4)$

c) $3^{-4} \cdot 3^{-2}$ d) $2^5 + 2^3$

2.3 POLYGON PERIMETER PATTERN

SP-27.   **Polygon Perimeters** The perimeter (distance around) of the polygon in Figure 1 is eight
units and the perimeter of the polygon in Figure 2 is 14 units.

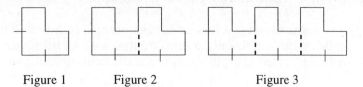

Figure 1   Figure 2   Figure 3

a)   Find the perimeter of the polygon in Figure 3. Find the perimeter of the polygon that
would be drawn for Figure 4.

b)   If the pattern of forming polygons continues as shown above, what would the perimeter
of the polygon in Figure 7 be? What would the perimeter of the polygon in Figure 9 be?
(Hint: Make a table of data and look for a pattern.)

c)   Explain in a sentence or two how to determine the perimeter of any polygon in this
pattern.

d)   If the figure number is N, state the formula (rule) for finding the perimeter, P, of the
polygon in Figure N.

SP-28.   a)   Fill in the data table from the Polygon Perimeters problem for at least ten figures. Copy
the axes on your paper. Then use the data to make a graph. You will need to plan ahead
and decide how to scale the axes.

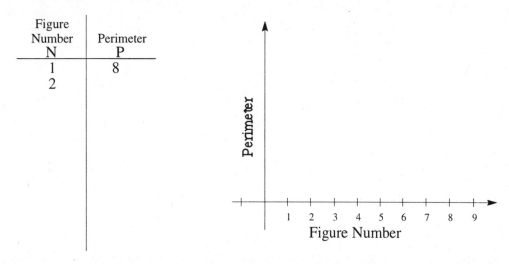

b)   Estimate the coordinates of the point where the pattern formed by the points would cross
the vertical axis. Which Polygon Perimeters figure would this point represent? Is it
possible to draw this figure? Explain.

SP-29.   a)   Should the points on your Polygon Perimeters graph be connected? Or, are there points (N, P) that don't make sense in the context of the problem?

b)   Read the following descriptions and copy the information in your Tool Kit.

Descriptions: continuous graphs and **discrete** graphs

A graph is **continuous** if it has no breaks, jumps, or holes. The points you get from a table can be connected to form a line or curve without lifting the pencil from the paper. Mathematicians often assume a graph is continuous unless a specific context implies it is not.

A graph is **discrete** if each point on the graph is separated from the other points on the graph. In the context of a problem, it won't make sense to connect the points (x, y) you get from a table for the rule.

c)   Should the points on your graph of Ms. Speedi's commute data be connected? Or, are there points (x, y) that don't make sense in the context of the problem? Answer these questions and use your answers to explain whether the graph is discrete or continuous.

In problems SP-30 and SP-31, copy and complete each table, and write each rule in words.

SP-30.

IN (x)	OUT (y)
3	−6
7	−14
6	−12
4	
10	
−1	
100	
x	

SP-31.

IN (x)	OUT (y)
3	7
1	5
5	9
2	
	10
−10	
	21
1000	
x	

For each of problems SP-32 through SP-35:

   a)   Copy and complete the table.

   b)   Explain in words what is done to the input value, x, to produce the output value, y.

   c)   Write the process you described in part (b) in algebraic symbols.

   d)   Graph the data. Use the same scale on both the x-axis and the y-axis. (You may not be able to fit all the points from a table on your graph because their x or y values are too large or too small.)

SP-32.

IN (x)	9	−6	10		20	7	−2		x
OUT (y)	27	−18		15					

SP-33.

IN (x)	0	2	6	−8		−10	−28		x
OUT (y)	0	1	3	−4	2				

SP-34.

IN (x)	2	1	−3.5	8.1	5.8	0.3	−		x
OUT (y)	0.2	0.1	−0.35				−1		

SP-35.

IN (x)	5	10	$6\frac{1}{2}$	$15\frac{1}{2}$	−9				x
OUT (y)	$7\frac{1}{2}$	$12\frac{1}{2}$				$4\frac{1}{2}$	12		

SP-36.   Imagine you have two standard dice -- one is orange and the other is yellow.

   a)   If you roll the orange die, how many outcomes are possible?

   b)   If you roll the orange die, what is the probability of getting a 5 ?

   c)   If you roll both dice, how many outcomes are possible? (Hint: The answer is not 12.)

   d)   If you roll both dice, what is the probability of getting a total of 5 points?

SP-37.   Examine this pattern:

a)   Draw the eighth figure in the pattern.

b)   Draw the thirtieth figure in the pattern.

c)   Explain how to decide what the 366th figure would look like.

SP-38. $\frac{2}{5}$ $\frac{5}{2}$ **Reciprocals**   When Max was in fifth grade, he learned that to find the reciprocal of a number, you just "flip it over." He recently learned the following definition, which is more precise.

> Definition: **Reciprocal**
> Two numbers are **reciprocals** of each other if their product is one.
> For example, 2 and $\frac{1}{2}$ are reciprocals of each other, since $2 \cdot \frac{1}{2} = 1$.

a)   Use the precise definition to explain why $1\frac{1}{3}$ is the reciprocal of $\frac{3}{4}$.

b)   What is the reciprocal of $\frac{3}{4}$ using Max's method? Explain why Max's method works.

c)   Use the precise definition to explain why, for any number $x \neq 0$ you choose, $\frac{1}{x}$ is the reciprocal of x.

d)   Add the definition of reciprocal to your Tool Kit.

SP-39. $\boxed{\frac{1}{x}}$ or $\boxed{x^{-1}}$ **Calculating Reciprocals**   There are two ways to find the reciprocal of a number on your calculator.

Method 1: Use division.
        For example, to find the reciprocal of 2,
        press $\boxed{1}$, $\boxed{\div}$, $\boxed{2}$, $\boxed{=}$. The calculator display should be 0.5,
        which is the correct reciprocal since $(2)(0.5) = 1$.

Method 2: Use the $\boxed{\frac{1}{x}}$ or $\boxed{x^{-1}}$ key.

        For example, to find the reciprocal of 2,
        press $\boxed{2}$, then the $\boxed{\frac{1}{x}}$ or $\boxed{x^{-1}}$ key. The calculator display
        should be 0.5, which is the correct reciprocal since $(2)(0.5) = 1$.

[PROBLEM CONTINUED ON NEXT PAGE]

SP-39. continued

Update your Calculator Tool Kit, and then use your calculator to find the reciprocal of each of the following values of x.

a)   2   b) $-2\frac{1}{2}$ c)   0.5   d)   100

e)   −100   f)   0.01   g)   −0.75   h) $\frac{1}{2}$

i)   3   j) $\frac{2}{5}$ k)   −0.4   l) $\frac{9}{2}$

SP-40.   Copy each of the following problems and carry out the calculations.

a)   −427 + (−85.3)   b)   (−32)(−3)(−7)

c)   −17 − (−492)   d)   −17 ÷ (−0.05)

e) $-\frac{2}{3}+\frac{3}{4}$ f) $-1\frac{2}{3}+7$

g) $(4\frac{1}{5})(-8)$ h)   −7 + (−8)(−6)

i)   18 + (−9) ÷ (−3) + 72   j) $2(-3)^2$

SP-41.   Find four consecutive odd integers such that the sum of the second integer and twice the fourth integer is 65. Use a Guess and Check Table to solve this problem.

SP-42.   Walter says that $x^7 \cdot x^3 = x^{10}$. James says that $x^7 \cdot x^3 = x^{21}$. Who is correct? Write an explanation that will convince the person who is wrong why he is wrong.

SP-43.   Write each of the following exponential forms in both fraction form and decimal form.

a) 4^{-3} b) 7^0 c) 5^{-2} d) 8^{-1}

SP-44.   a)   Copy this sequence, fill in the blanks, and write the eighth line of the pattern without writing lines six and seven. A calculator will be helpful.

$$1 \cdot 9 + 2 = \underline{\hspace{1cm}}$$
$$12 \cdot 9 + 3 = \underline{\hspace{1cm}}$$
$$123 \cdot 9 + 4 = \underline{\hspace{1cm}}$$
$$\vdots \qquad\qquad \vdots$$
$$\underline{\hspace{1cm}} \cdot 9 + \underline{\hspace{0.3cm}} = \underline{\hspace{1cm}}$$

b)   How far will this pattern continue? Experiment to check whether your conjecture is true.

2.4 PUNCH MIX

SP-45.   Matthew was making fruit punch on a hot Sacramento Valley summer day. He started with three quarts of grape juice. He then added two quarts each of several different kinds of fruit juice. Eventually, he had 19 quarts of fruit punch.

a)   Make a table showing the relationship between the number of kinds of fruit juice (the input values, x) and the total amount of fruit punch after each addition (the output values, y) and then draw a graph of the relationship. Don't forget the grape juice!!

b)   How many kinds of fruit juice did Matthew use in making the 19 quarts of punch?

c)   If Matthew wanted to make 25 quarts of punch, how many kinds of fruit juice would he need?

d)   If Matthew used 15 kinds of juice, how much punch would he make?

e)   Write a rule (equation) that describes the relationship shown in the table and on the graph.

f)   Re-read the Punch Mix problem and then answer the following questions about your graph:

Should your graph show negative values for x ? for y ? Explain your answers.

Should the points on your graph be connected? Explain your answer.

SP-46.   In the Algebra Walk, you graphed the rule $y = x^2 - 1$. In this problem you'll examine the graph of the "squaring" rule, $y = x^2$. First, copy and complete this table for $y = x^2$:

x	−4	−3	−2	−1	−0.5	0	0.5	1	2	3	4
y											

Using the values from the table and additional inputs provided by your instructor, graph the equation $y = x^2$ with your group. Have one person place your special group points on the board graph. Then answer the following questions by discussing them with your group:

a)   Is this graph a straight line?

b)   Connect the points with a smooth curve and describe the result.

c)   Can any value for x be used? Explain.

d)   Can any y-value result from this rule? Explain.

e)   Use the graph to approximate the x-value(s) that correspond to $y = 5$.

f)   Use your calculator to solve part (e) more precisely.

SP-47.   Write an expression for the length of each composite line segment. For example,

the segment  has a length of x units, while

the composite segment ⊢ a ⊢ a ⊢ 5 ⊣ has a length of a + a + 5, or 2a + 5 units.

a) ⊢——— q ———⊢——— 3m ———⊣

b) ⊢ m ⊢ m ⊢ m ⊢——— 5m ———⊣

c) ⊢ x ⊢ 2x ⊢——— 7 ———⊣

d) ⊢ y ⊢ y ⊢——— 11 ———⊢ y ⊣

e) ⊢ n ⊢ m ⊣

SP-48.   Suppose you randomly toss four coins, a penny, a nickel, a dime, and a quarter. They might come up "heads, heads, heads, heads" (H, H, H, H).

a)   What other possible outcomes might occur? Describe the strategy you will use to make sure you list all the possibilities, including "H, H, H, H," and then list them all.

b)   Find the probability that exactly one coin comes up "heads."

c)   Find the probability that at most two coins come up "heads."

In each of problems SP-49 through SP-51:

a)   Copy and complete each table.

b)   Write a rule using the variable x.

c)   Next, graph the rule and estimate the coordinates of the point where the graph crosses the x-axis. Be sure to label your graph and the axes.

SP-49.

x	1	10	5	0	−1		2		x
y	−2	7	2			3		−7	

SP-50.

x	2	−1	0	−2	$\frac{1}{2}$			x
y	−4	2	0			6	3	

SP-51.

x	2	10	6	7	−3		−10	100	x
y	8	32	20		−13				

SP-52.   Compare the graphs you made in SP-49 and SP-51. How are they alike and how are they different?

SP-53.   Here's a shading pattern similar to the one in SP-10:

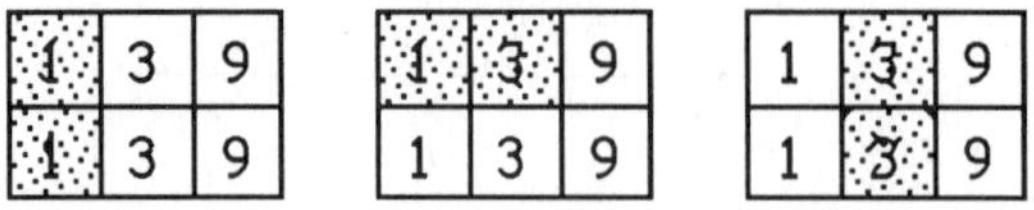

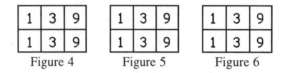

Figure 1   Figure 2   Figure 3

a)   Copy the three unshaded figures and continue the pattern by shading them.

1	3	9
1	3	9

Figure 4

1	3	9
1	3	9

Figure 5

1	3	9
1	3	9

Figure 6

b)   Write a sentence that states a rule for your pattern.

c)   Draw another unshaded figure like those above and shade it to represent Figure 11 in your pattern.

d)   What is the largest number you can represent by shading one of the figures?

SP-54.   Write the rule $y = 3x^2$ in everyday words.

SP-55.   Multiply 583.85 by 1 billion. Write the product in scientific notation.

SP-56.   Yu-Ping kept track of her checking account balance at the beginning of each month. Can she use this data to draw a continuous graph? Explain your answer.

SP-57.   Write 2^5 in six different ways. For example, $2^5 = 2^2 \cdot 2^3$.

SP-58.   Use a Guess and Check Table to solve the following problem. State your solution in a sentence.

Find two consecutive whole numbers such that the sum of their squares is 265.

2.5 SLICING A PIZZA

SP-59.   ◌̵ **Slicing a Pizza** In this problem you'll explore the following question about slicing a round pizza into pieces by making straight cuts through points on the edge of the crust:

> Suppose you have a perfectly round (circular) pizza. Around the edge of the crust are some olives. (There are no other olives on the pizza.). If you slice the pizza by making cuts that connect the olives, so that every pair of olives is connected by a cut, how many pieces of pizza will you make?

To explore this problem, you'll consider a succession of circles with points on them. Each successive circle will have one more point on it than the previous circle. To cut each circle into pieces, you will connect each possible pair of points with a line segment.

a)   On your copy of the resource page for SP-59, mark one point on a circle, as in Figure 1.

On another circle, add a second point and connect the points as in Figure 2. On a third circle, mark three points and connect pairs of points as in Figure 3. Repeat the process with a fourth circle as in Figure 4.

Now count the number of pieces of pizza (regions) in each figure. For example, Figure 3 has four regions.

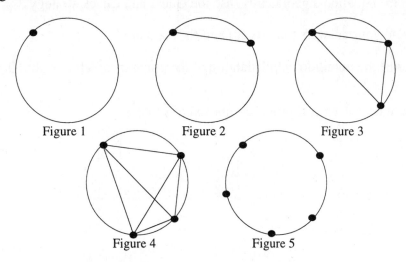

Figure 1   Figure 2   Figure 3

Figure 4   Figure 5

b)   Complete the fifth circle in the pattern and count the regions.

c)   Make a table to organize the data you've collected so far.

d)   Suppose a circle had six points marked on it and you connected each pair of points with a line segment. Use your data to <u>predict</u> the number of regions that would be formed in the circle.

e)   Test your prediction from part (d) by completing an appropriate figure on your resource page and then counting the regions.

SP-60.   Write a short paragraph to describe what you did and your results in problem SP-59. What did the problem help you understand about looking for and finding patterns?

SP-61.   a)   Copy and complete this table:

x	2	0	-3	$\frac{1}{2}$	-1	0.3		$-\frac{1}{3}$		-11	-2	x
y	5	1	-5	2			7		10			

b)   Describe in words the rule you used in part (a).

c)   State the rule using a variable such as x.

d)   Graph the ordered pairs (x, y).

e)   Find three more pairs of points that satisfy the rule without using any more <u>whole</u> numbers.

f)   If everyone on campus today used three different x-values, found the corresponding y-values, and put these points on the same graph, would the "line" formed by the points be complete?

g)   Estimate where the graph crosses the x-axis.

h)   Find a value for x that makes the equation $2x + 1 = 0$ true. If you are unsure how to solve the equation algebraically, use the Guess and Check strategy.

i)   How do your answers to parts (g) and (h) compare?

j)   Describe the uphill/downhill slanting of the graph from left to right. Is it very steep?

SP-62.   In the pattern below, each figure is composed of squares.

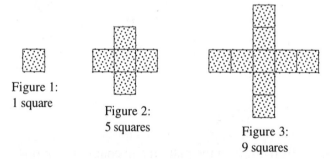

Figure 1:
1 square

Figure 2:
5 squares

Figure 3:
9 squares

a)   Determine how many squares Figure 4 will have.

b)   Record all of your information in a table. Use a pattern you see in your table to predict how many squares there will be in Figure 10. How many squares do you think there will be in Figure 100 ? Write one or two sentences to explain how to find the number of squares for any figure in this pattern.

c)   Suppose you graphed the points in your table with the Figure Number on the horizontal axis, and the Number of Squares on the vertical axis. Do you think the graph would be linear, or not? Explain your answer.

d)   If you graphed the points in your table as described in part (c), should your graph be discrete or continuous? Explain.

SP-63. **Symbols**: " ≤ " and " ≥ "
The symbol " ≤ " means "less than or equal to."
For example, $x \le -5$ says "x is less than or equal to –5."

The symbol " ≥ " means "greater than or equal to."
For example, $x \ge 7$ says "x is greater than or equal to 7."

We can express the idea that
"x is greater than or equal to –3 and x is less than or equal to 7"
by writing
$$-3 \le x \le 7.$$
This means that x can be any number between –3 and 7, including –3 and 7. For example, x could be –2 because $-3 \le -2 \le 7$; or x could be 1.05 because $-3 \le 1.05 \le 7$.

a) Copy the information about the "≤" and "≥" symbols in your Tool Kit.

b) Suppose x must fit the restriction: $-1 \le x \le 2$. List ten different numbers which meet this constraint.

c) Suppose x must fit the restriction: $x \ge 5$. List five different integers, and five numbers which are not integers, that meet this constraint.

SP-64. Joyce and John each have a number. The sum of these numbers is fifteen.

a) Make a table showing several pairs of possible numbers and graph your results.

b) If John's number is $9\frac{1}{3}$, what is Joyce's number?

c) Should the points on your graph be connected to form a line? Explain.

SP-65. The length of a rectangle is three times the width. If the area is 18.75, find the width and length. Use a Guess and Check Table to solve the problem.

SP-66. Write the meaning of each exponent expression and **then** simplify using exponents. For example,
$$(x^3)^2 = (x^3)(x^3) = (x \cdot x \cdot x) \cdot (x \cdot x \cdot x) = x^6.$$

a) $(x^4)^2$

b) $(x^2)^3$

c) $(x^5)^5$

d) $(x \cdot y)^2$

e) $(x^2 \cdot y^3)^3$

f) $(2x)^4$

SP-67. Copy and solve these Diamond Problems:

2.6 EGG TOSS

SP-68.   ☺ **Egg Toss** If an egg is tossed into the air at a rate of 80 feet per second, its height y above the ground (in feet) after x seconds is given by $y = 80x - 16x^2$. (This is a formula used in physics.)

   a)   Make a table for $0 \le x \le 5$. Be sure to include some non-integer values such as 0.25, 0.5, 4.8, etc.

   b)   Graph your information from part (a). Scale the x-axis in increments of one-half and the y-axis in increments of five.

   c)   How high is the egg after one second?

   d)   When is the egg 96 feet high?

   e)   When is the egg 200 feet high?

   f)   Describe the shape of the graph and what it represents in terms of tossing an egg into the air.

   g)   When does the egg reach its maximum height? What is the maximum height the egg reaches?

   h)   For what values of x is $y = 0$? What does this represent about the egg?

   i)   If $x = 6$, what is the value of y? What might this number represent?

SP-69.   Scott says that $-x^2$ is the same as $(-x)^2$, but Ximena insists that they are different.

   a)   The first expression, $-x^2$, says, "Square x and find the opposite (negative) of the product." Try using these directions to evaluate $-x^2$ for the values $x = -3, -1, 0, 2$. Make a table of your results:

x	−3	−1	0	2
$-x^2$				

   b)   The second expression, $(-x)^2$, says, "Find the opposite of x, **then** square it." Try using these directions to evaluate $(-x)^2$ for the values $x = -3, -1, 0, 2$. Make a table of your results.

   c)   Which claim is correct? Convince Scott and Ximena by explaining why.

SP-70.   Rowan and Martin made a careful graph of $y = \sqrt{x}$ in their algebra class last week.

a)   When Rowan made a table of values, the first value he chose for x was 9. Find the corresponding value of y, and label that point on the graph with its coordinates.

b)   Without using a calculator, use the graph to estimate the value of ? that will make the equation true: $\sqrt{3} = ?$

c)   Without using a calculator, use the graph to estimate the value of ? that will make the equation true: $\sqrt{?} = 5$

d)   Without using a calculator, use the graph to estimate the value of ? that will make the equation true: $\sqrt{?} = 4.5$

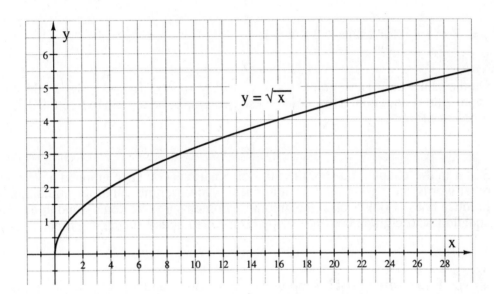

SP-71.   Sketch a graph to represent each of the situations or relationships described below. Label the axes. The vertical axis is given first.

a)   the total cost of gasoline compared to the number of gallons of gasoline you buy

b)   the amount of gasoline in your tank compared to the number of hours you have driven your car since you filled the tank

c)   the height of a burning candle compared to time

d)   the height of an unlit candle compared to time

SP-72.   **Without** drawing a graph, describe the shape of the graph of each of the equations below.

a) $y = x^2 - 3$

b) $y = 2x - 3$

For each of problems SP-73 and SP-74:

    a)   Make a table using several values, including fractions and decimals, in the given domains (sets of input values).

    b)   Graph the points you found in part (a). Be sure to label the appropriate parts of each graph.

    c)   Use the graph to approximate the x-value that corresponds to a y-value of 1.

    d)   Determine the points (x, y) where the graph crosses the x-axis.

SP-73. $y = x^2 - 4$ for values $-3 \le x \le 3$

SP-74. $y = -x^2 + 4$ for values $-3 \le x \le 3$

SP-75.   On the same set of axes, graph $y = -x^2$ and $y = (-x)^2$ for $-3 \le x \le 3$. Be sure to label each gaph with its equation. How could these two graphs be of help to Scott and Ximena in problem SP-69 ?

SP-76.   So far in this chapter, most of the rules or equations you have graphed have been **linear** (in other words, their graphs are straight lines). However, not all of the equations you graph will produce lines. Look back at the graphing problems you've done. Find three examples of an equation whose graph is a line. Copy each equation and sketch its graph.

SP-77.   **Pair-a-dice Lost?** Suppose you have two standard six-sided dice. If both dice are rolled, how many outcomes are possible?
How many ways are there of rolling a total of six on these two dice?
What is the probability of rolling a total of six ?

SP-78.   Use your calculator's y^x or x^y key to compute each of the following values.

    a) 2^3 b) 3^0

    c) 2^{-1} d) $(\sqrt{3})^4$

SP-79.   a)   Write the meaning of the expression $x^{-4}(x^2)^3$, and then write it in simplified exponential form.

    b)   Write the meaning of the expression $5(2x^3y^2)^4$, and then write it in simplified exponential form.

SP-80.   Copy and solve these Diamond Problems:

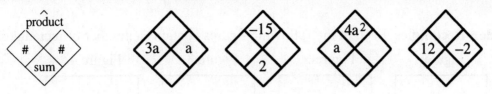

SP-81.   Solve this problem by using a Guess and Check Table:

Find two consecutive integers such that the sum of the first integer and five times the second integer is 107.

2.7 COUNTING RECTANGLES AND FOLDING BILLS

SP-82.   **Hidden Rectangles** Each of the following figures contains a number of rectangles:

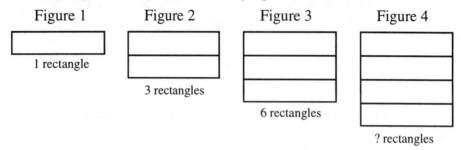

a)   How many rectangles are in the fourth figure?

b)   How many rectangles would be in the fifth figure?

c)   How many rectangles would be in the sixth figure?

d)   Without drawing any more figures, determine how many rectangles would be in the tenth figure and explain how you solved the problem. A table may be helpful.

SP-83.   a)   Use the information you found in problem SP-82 to complete the following table for at least ten figures.

Figure Number, N	Number of Rectangles, R
1	
2	
3	
⋮	
10	

b)   Graph the information from your table using a grid with the Figure Number, N, on the horizontal axis, and the Number of Rectangles, R, on the vertical axis. Plan your scale so that all ten points will fit on your graph. Label the scale and the axes.

c)   Should your graph have any negative values for N ? for R ? Explain your answers.

d)   Should your graph be continuous or discrete? Explain your answer, and then decide whether you should connect the points on your graph.

SP-84. ▷ **Dollar Bill-Folds** You found in OP-45 that a one-dollar bill is about 0.011 cm thick. Suppose you folded it in half once -- how thick would the two layers be? If you folded the dollar bill in half again, you'd have four layers. Imagine you keep folding the dollar bill in half to make more and more layers.

a) First, make a prediction. How thick do you think the folded dollar bill will be after 50 folds?

b) Make a table to record the results of your bill-folding with "number of folds" as inputs and "thickness" of layers as outputs. Do this for up to ten folds.

Number of folds, F	0	1	2	3	4	...	10
Thickness, T, in cm	0.011	(0.011)(2) = 0.022				...	

c) Plot the information from your table on a grid, with "number of folds" on the horizontal axis and "thickness" on the vertical axis. Does it make sense to connect the points on your graph? Explain.

d) Describe in a few sentences a pattern that will allow you to find the thickness of the dollar bill after a given number of folds.

e) *Really stretch* your imagination and suppose you could fold a dollar bill in half 50 times! Without filling in the rest of the table, determine how thick the dollar bill would be after 50 folds. Give this thickness in centimeters and kilometers. (There are 100 cm in one meter, and 1000 meters in one kilometer.)

SP-85. **Symbol: "..."**
The symbol "..." is called an **ellipsis**. It indicates that certain values in an established pattern, or sequence, have not been shown, although they are part of the sequence.
For example, in the sequence
$$3, 3.5, 4, \ldots, 8.5, 9$$
the ellipsis (...) means that the values 4.5, 5, 5.5, 6, 6.5, 7, 7.5, and 8 are part of the sequence even though they are not written.

a) Copy the information about "..." in your Tool Kit.

b) List all of the numbers in the sequence given by 8, 8.2, 8.4, 8.6, ..., 9.8, 10.

c) List all of the numbers in the sequence given by −4, −3.5, −3, ..., 2.5, 3.

SP-86. For each equation below, make a table using the input values provided. Then graph the equation and label your graph. For part (c), the table is already given.

a) $y = 3x^2 - 5$ for $x = -2, -1.5, -1, \ldots, 1.5, 2$

b) $y = -x^2 + 2x - 3$ for $x = -2, -1, 0, \ldots, 3, 4$

c) $y = x^2 + 3x + 1$ for $x = -4, -3, -2, \ldots, 1, 2$

x	−4	−3	−2	−1	0	1	2
y	5	1	−1	−1	1	5	11

For part (c), there is one more x-value that is important to the shape of this graph--what is it? Why is this x-value important? Be sure to plot the point associated with this x-value on your graph.

SP-87.   Use your calculator to compute each of the following values.

a) 3^2 c) 0.5^0 e) 20^{-2}

b) 0.5^3 d) $1.5^{3.2}$ f) 2^{-3}

SP-88.   Robbie set up this Guess and Check Table for a problem about a rectangle:

Guess Width	Length	Perimeter	Check 60 ?
5	11	32	too low
7	15		
10	21		

a)   Copy and complete the table, as it is shown, but do not solve the problem.

b)   If the next guess for the width is 9, describe how you would calculate the length.

c)   Use your description from part (b) to write the length in symbols if the next guess for the width is x.

SP-89.   Copy the table below. Then rewrite each expression using exactly one set of parentheses in order first to maximize and then to minimize its value.

Expression	Greatest Value	Least Value
$2^2 \cdot 3^2 - 1^2$		
$5 + 7 \cdot 3^2 - 4$		
$4^2 \cdot 3 - 2^2 \cdot 5$		

SP-90.   Patty's Hamburgers advertises that the company has sold an average of 405,693 hamburgers per day for the past 15 years.

a)   About how many hamburgers total is this? Round your answer to the nearest 100,000.

b)   Express your answer in part (a) in scientific notation.

SP-91.   Write the meaning of each expression below and **then** simplify using exponents. Write each answer in exponential form, if possible; otherwise, write it in standard form.

a) $3^5 \cdot 3^2$ b) $3^5 + 3^2$ c) $3^5 \div 3^2$ d) $3^5 - 3^2$

SP-92.   a)   Ms. Escargot drove 50 mph for 45 minutes. How far did she drive? Show how you get your answer.

b)   Mr. Snail drove 25 mph for 80 miles. How long did it take him? Show how you get your answer.

SP-93.   Sally says $10^3 \cdot 10^4 = 100^7$, but Elaine is sure that $10^3 \cdot 10^4 = 10^7$. Which calculation is correct? Use examples or numbers to explain your reasons.

SP-94.   Find the perimeter of Mr. Escargot's garden.

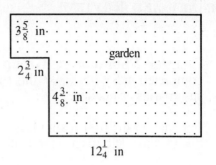

SP-95.   **Chapter 2 Summary: Rough Draft** You've been doing lots of problems. Now it's time to pause and to reflect on what the main ideas of this chapter are. By taking the time to carefully think and write about the focal ideas of the chapter, you are helping organize your ideas about what you've learned in a way that should be meaningful and useful to you.

In Chapter 2 you've learned an important problem solving strategy: to look for, describe and use patterns. You've used patterns to make conjectures, and have seen that it is important to test conjectures. You have also learned to apply your patterning skills to make graphs of rules.

The main ideas of Chapter 2 are:

*   An important problem solving strategy is to look for, describe, and use patterns. We can use patterns to make conjectures. It is important to test conjectures.

*   Graphs, tables, and rules are related.

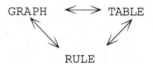

*   The graph of an equation follows a pattern. A graph can be **discrete** (made of individual points) or **continuous** (made of connected points) or neither.

    A point on a graph that doesn't follow the pattern of the other points is probably miscalculated.

*   The graphs of some equations are **lines**. Other rules produce **parabolas** as their graphs. Still other equations produce **curves** that are not necessarily parabolas.

Write your answers to the following questions in rough draft form **on separate sheets of paper**, and be ready to discuss them with your group at the next class meeting. Focus on the **content**, not neatness or appearance, as you write your summary. You will have the chance to revise your work after discussing the rough draft with your group.

a)   For each of these four ideas, select and neatly recopy a problem that you did well, and write why you think that the selected problem is a good representative of the idea.

b)   In this chapter you have looked at the graphs of many equations. Write a paragraph to describe relationships you've noticed between equations and graphs. You may wish to include an example to illustrate your ideas.

c)   What were the most difficult parts of this chapter? List sample problems and discuss the hard parts.

d)   What problem did you like best and what did you like about it?

2.8 SUMMARY AND REVIEW

SP-96.   **Chapter 2 Summary: Group Discussion** Take out the rough draft summary you
           completed in SP-95. Take this time to discuss your work; use homework time to revise your
           summaries as needed.

           For each of the four main ideas of the chapter, choose one member of the group to lead a short
           discussion. The discussion leaders should take turns to:

              • explain the problem they chose to illustrate their main idea,

              • explain why they chose that particular problem,

              • tell which problem they liked best and what they liked about it, and

              • tell what they thought were the most difficult parts of this chapter.

           This is your chance to make sure your summary is complete, update your Tool Kits, and work
           together on problems you may not be able to solve yet.

SP-97.   a)   Explain how to enter $\frac{12}{3}$ on your calculator. Do it.

           b)   Try $\frac{0}{1}$ on your calculator. What happens?

           c)   Try $\frac{1}{0}$ on your calculator. What happens?

SP-98.   a)   Graph $y = \frac{1}{x}$ for $x = -5, -3, -2, -1.5, -1, -0.5, -0.33, -0.2, 0, 0.2, 0.33, 0.5, 1,$
                1.5, 2, 3, and 5. Scale the axes so that two marks on the graph paper represent one unit.

           b)   For positive values, what happens to y as x gets larger? What happens to y as x
                gets smaller?

           c)   What happens to y when x decreases from 1 to 0 ? What happens to y when x
                increases from –1 to 0 ?

           d)   Use the graph to <u>estimate</u> a value for x so that $\frac{1}{x}$ is ...

                    1)   2   2) $\frac{1}{2}$

                    3)   –2
                                                             4)   0.3
                    5) $\frac{-1}{4}$
                                                             6)   2.5

SP-99.   The Happy Lunch Shop sells four kinds of sandwiches and three kinds of drinks. How many
           different combinations of one sandwich and one drink could a customer choose for lunch? (To
           be sure you've counted all possiblities, make a list.).

SP-100.   Use a Guess and Check Table to solve the problem below.

Eduardo used raspberry juice and lemon-lime soda to make $11\frac{1}{2}$ quarts of punch for his party. If there were four more quarts of soda than raspberry juice, how much of each beverage was used in the punch?

For each equation in problems SP-101 and SP-102, make a table of values and graph the equation.

SP-101. $y = x^3$ for x = −2, −1.5, −1, ... , 1.5, 2

SP-102. $y = x$ for x = −4, −3, ... , 3, 4

SP-103.   Hakeem set up this Guess and Check Table for a problem about money:

Guess # nickels	# dimes	# quarters	value nickels	value dimes	value quarters	Total	Check = $ 3.55 ?
5	7	4	$0.25	$0.70	$1.00	$1.95	too low
7	9	6					
12	14	11					

a)   Complete the table as it is shown, but do not solve it.

b)   If the next guess for the number of nickels is 10, describe how you would calculate the number of dimes.

c)   If the next guess for the number of nickels is 10, describe how you would calculate the value of the quarters.

d)   If the next guess for the number of nickels is x, write your descriptions from parts (b) and (c) in symbols.

SP-104.   Compute the numerical value of each of the five-digit expressions below.

a) $3 + 3 \div (3 + 3) + 3$ b) $3 \cdot 3 + 3^3 \div 3$

SP-105.   Use the graph to ...

a)   complete a table of values for the points,

b)   and find the rule y = _____.

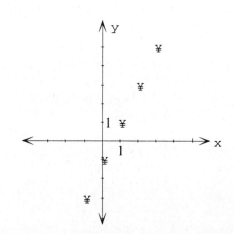

SP-106.   For each of the following rules, copy and complete the given table.
           Then use the table to graph the rule.

   a) $y = -x + 4$

x	−4	−3	$-\frac{1}{2}$	0	$\frac{1}{2}$	0.3	1	2
y								

   b) $y = x^2 - 2$

x	−3	−2	−1	−0.5	0	1	1.3
y							

   c) $y = x^2 + 2x + 1$

x	−3	−2	−1	0	1
y					

   d) $y = 2^x$

x	−2	−1	−0.5	0	0.5	1	2
y							

SP-107.   Suppose that your grandmother gave you $1 on your first birthday, $2 on your second
           birthday, $3 on your third birthday, etc. Also suppose that you saved **all** the birthday money
           your grandmother gave you in a piggy bank and never took any money out.

   a)   Let your age in years be the input (x), and let the total savings from all your birthdays be
         the output (y). Organize your data in a table and draw a graph that includes at least your
         first four birthdays.

   b)   Describe the graph and explain what it represents in two or three sentences.

   c)   Are there any points on your graph <u>between</u> years one and two? Explain.

   d)   How much birthday money will you have in your piggy bank on your sixteenth birthday?

   e)   How much birthday money will you have altogether on your twenty-first birthday?

SP-108.   **Chapter 2 Summary: Revision** This is the final summary problem for Chapter 2. Using
           your rough draft from SP-95 and the ideas you discussed in your groups from SP-96, spend
           time revising and refining your Chapter 2 Summary. Your presentation should be thorough
           and organized, and should be done on a separate piece of paper.

2.9 THE BURNING CANDLE INVESTIGATION (OPTIONAL)

THE BURNING
CANDLE

SP-109.   **The Burning Candle Investigation** Suppose it's your friend's birthday and you want to surprise her by walking into the room carrying a piece of cake with a lighted candle. Could you predict how long before the candle goes out?

To answer this question, you'll use a video presentation of a burning candle to collect data and then make a graph and look for a pattern.

a)   Gather data from the "Burning Candle" video presentation. Note the mass of the candle at various times during the presentation and write down your observations. You should make <u>at least</u> five observations, with at least 40 seconds between observations. Be sure to write down both the time and the associated candle mass.

b)   Make a table from your data.

c)   Make a new column in your table to show elapsed time.

d)   Set up a graph (using equal intervals) with elapsed time on the horizontal axis.

e)   Graph your data, comparing mass, m, to elapsed time, t.

f)   Sketch a line or curve connecting your data points. In your group, compare graphs and decide which is the most accurate. Copy your group's choice.

g)   Use your group's graph to predict the mass of the candle at the elapsed time of 1:20. Check the accuracy of your group's prediction by reviewing the "burning candle" video.

h)   Use your group's graph to predict the mass of the candle at the elapsed time of 2:47. Check the accuracy of your group's prediction by reviewing the "burning candle" video.

SP-110.   Use your graph form the Burning Candle Investigation to answer this question:

"If the candle continued to burn, when do you think it would go out?"

After discussing the question with your group, carefully explain your answer in complete sentences.

SP-1. Algebra Walk Observations

Sketch what you see: Describe what you see, and write the rule in words:

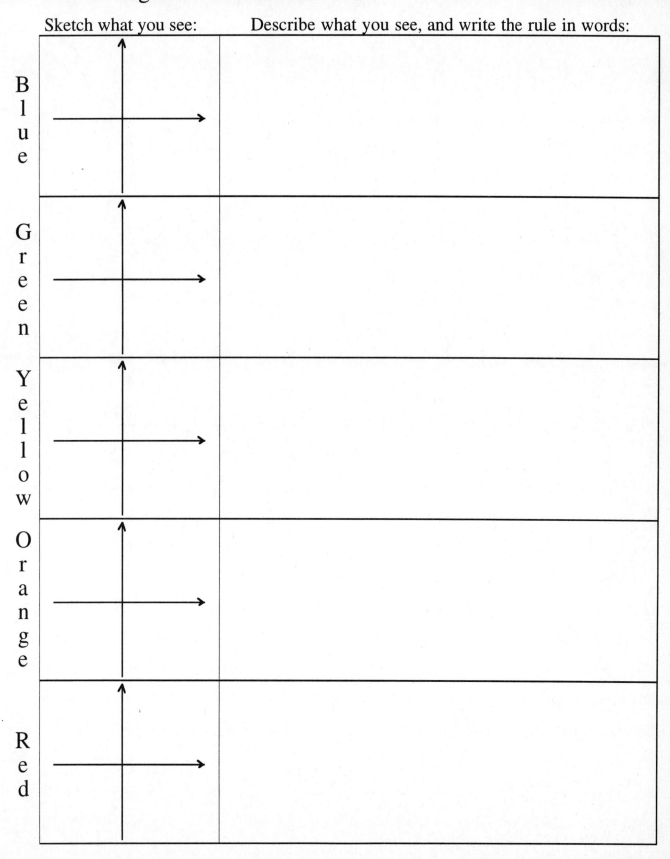

GRID SHEET

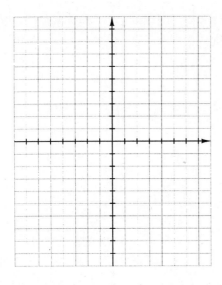

IN (x)										
OUT (y)										

IN (x)										
OUT (y)										

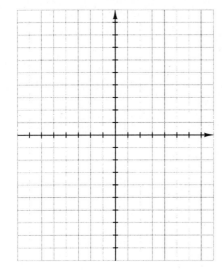

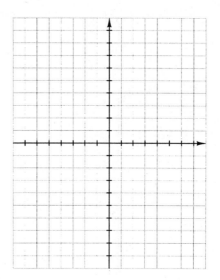

IN (x)										
OUT (y)										

IN (x)										
OUT (y)										

GRID SHEET

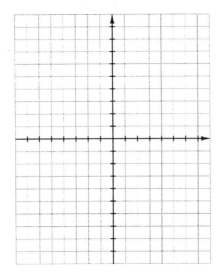

IN (x)									
OUT (y)									

IN (x)									
OUT (y)									

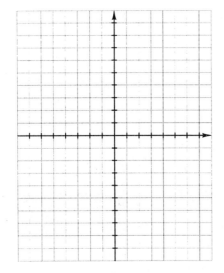

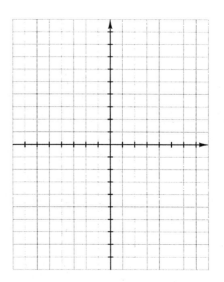

IN (x)									
OUT (y)									

IN (x)									
OUT (y)									

Algebra Tool Kit: the "what to do when you don't remember what to do" kit

Algebra Tool Kit: the "what to do when you don't remember what to do" kit

Calculator Tool Kit: for instructions on using your calculator

Calculator Tool Kit: for instructions on using your calculator

SP-59. Slicing a Pizza
Suppose you have a perfectly round (circular) pizza. Around the edge of the crust are some olives. (There are no other olives on the pizza.). If you slice the pizza by making cuts that connect the olives, so that every pair of olives is connected by a cut, how many pieces of pizza will you make?

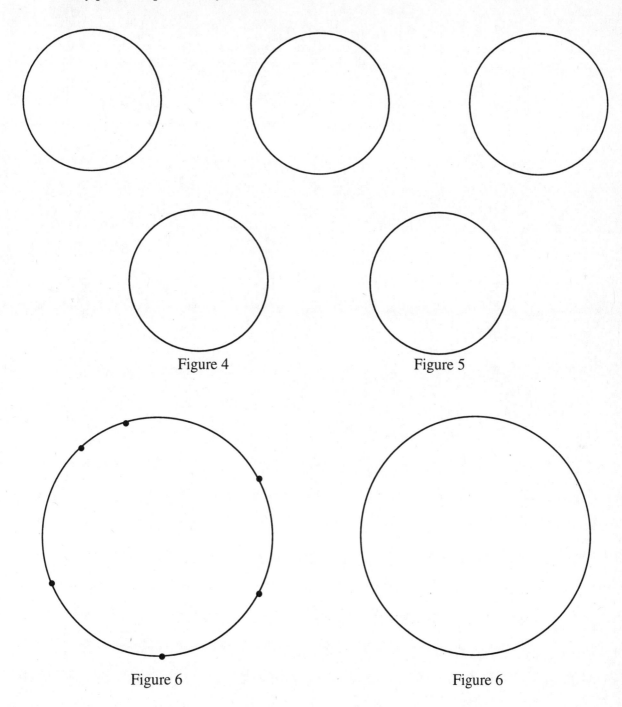

Figure 4   Figure 5

Figure 6   Figure 6

Chapter 3

CHOOSING A PHONE PLAN: Writing and Solving Equations

$$3c + 1 = -8$$

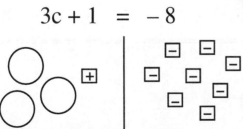

CHAPTER 3

CHOOSING A PHONE PLAN:
WRITING AND SOLVING EQUATIONS

This chapter is a powerful one. In it you will tie together the organizational and patterning skills you developed using Guess and Check Tables and the process of writing algebraic equations to represent word problems.

The power of knowing how to solve equations lies in the relative ease and efficiency that the algebraic approach gives you. However, it may be difficult to figure out, in the case of word problems, for example, exactly which equation it is that you should solve.

The power of a Guess and Check Table, on the other hand, is that it helps you (through organization and patterning) to write an equation. You can also use Guess and Check Tables to solve problems. However, many problems are extremely tedious, if not impossible, to solve this way.

In this chapter you will have the opportunity to:

- write equations from Guess and Check Tables;

- use the distributive property to rewrite expressions; and

- solve linear equations.

CHAPTER CONTENTS

3.1 WRITING EQUATIONS FROM GUESS AND CHECK TABLES

Many situations in daily life, like scheduling deliveries, ordering supplies, and calculating costs of health insurance or phone service plans, can be represented by linear equations. The ability to write and solve equations associated with everyday problems is a useful, timesaving skill. Many of you have experienced a situation similar to the following one:

> In a college dormitory, each student has a choice of two phone companies. Company A charges $7.46 per month plus 13 cents a call, while Company B charges $6.17 per month plus 17 cents per call.

Which phone company would be the best choice for you? What would you do to help make your decision?

You already know you can investigate this problem by Guessing and Checking. In this chapter you will learn to use a more efficient way to solve problems. The skills you have developed using a Guess and Check Table to solve a problem will help you write equations for word problems, so the amount of guessing you must do is reduced to a minimum. Then you'll learn how to solve equations which are "linear."

Problems that are especially important are marked with a ⚷ symbol. They help you develop understanding and/or consolidate ideas. Pay careful attention to these problems, and be sure to revise your work if necessary.

CP-1. You know how to use a Guess and Check Table to solve word problems. In this problem you'll see how to extend the use of a Guess and Check Table to include writing an algebraic equation to represent a word problem.

Copy the following example in your notebook.

<u>Example of Using a Guess and Check Table to Solve a Problem</u>

The length of a rectangle is three centimeters more than twice the width. The perimeter is 45 centimeters. Use a Guess and Check Table to find how long and how wide the rectangle is, and write an equation from the pattern developed in the table.

STEP 1   In Chapter 1 we built the following Guess and Check Table:

Guess Width	Length	Perimeter calculations	Check Perimeter = 45 ?
10	23	2·10 + 2·23 = 66	too high
5	13	2·5 + 2·13 = 36	too low
7	17	2·7 + 2·17 = 48	too high
6	15	2·6 + 2·15 = 42	too low
6.5	16	2·6.5 + 2·16 = 45	correct

STEP 2   Now what if we make W our next guess? What would the length be?

Guess Width	Length	Perimeter calculations	Check Perimeter = 45 ?
W	2W+ 3		

STEP 3   What would the perimeter be?

Guess Width	Length	Perimeter calculations	Check Perimeter = 45 ?
W	2W + 3	2·W + 2·(2W + 3)	

STEP 4   How do we want the perimeter to relate to 45 ?

Guess Width	Length	Perimeter calculations	Check Perimeter = 45 ?
W	2W + 3	2·W + 2·(2W + 3) = 45	

STEP 5   Write the equation generated in the table: $2·W + 2·(2W + 3) = 45$.

You'll work on techniques for solving this equation later in the chapter. The solution you find algebraically will correspond to the correct guess in the Guess and Check Table.

Solve each of problems CP-2 through CP-6 by using a Guess and Check Table. Use the pattern you develop in the table to write an equation. Write your solution in a sentence.

CP-2. Admission to the fair is $1 for children and $2 for adults. On Monday there were 80 more children tickets sold than adult tickets. Total ticket sales for Monday came to $980. How many of each type of ticket was sold?

CP-3. One number is five more than a second number. The product of the numbers is 3300. What are the numbers?

CP-4. Heather has twice as many dimes as nickels and two more quarters than nickels. The value of the coins is $5.50. How many quarters does she have?

CP-5. Find three consecutive numbers whose sum is 36.

CP-6. The length of Linnea's rectangular garden is two meters longer than twice its width. The perimeter is 46 meters. Find the dimensions of the garden.

Solve each of problems CP-7 through CP-10 by making a Guess and Check Table and then write an equation from your table. Write your solution in a sentence.

CP-7. At a football game 2000 tickets were sold. General public tickets sold for $7.50 and student tickets for $5.00. The total revenue was $11,625. How many student tickets were sold?

CP-8. Mary sold 105 tickets for the basketball game. Each adult ticket costs $2.50 and each student ticket costs $1.10. Mary collected $221.90. How many of each kind of ticket did she sell?

CP-9. A rectangular goat pen is enclosed by a barn on one side and by a total of 100 feet of fence on the three other sides. The area of the pen is 912 square feet. Draw a diagram of the goat pen and then find its dimensions.

CP-10. Chris is three years older than David. David is twice as old as Rick. The sum of Rick's age and David's age is 81. How old is Rick?

Solve each of problems CP-11 through CP-14 by making a Guess and Check Table and then write an equation from your table. Write your solution in a sentence.

CP-11. On a 520-mile trip, Chloë and Maude shared the driving. Chloë drove 80 miles more than Maude drove. How far did each person drive?

CP-12.   A notebook costs $0.15 more than a ball-point pen. The total cost of the pen and notebook is $4.05. How much does the pen cost?

CP-13.   Joe has twice as many dimes as nickels and has 15 coins total. How many of each coin does Joe have?

CP-14.   A rectangular sign is twice as long as it is wide. Its area is 450 square centimeters. What are the dimensions of the sign?

CP-15.   Use the meaning of exponents to rewrite each of the following expressions more simply.

   a) $4^3 \cdot 4^2$ d) $x^3 y^2 \cdot x^5$

   b) $4^2 \cdot 4^6 \cdot 7^3 \cdot 7^{10}$ e) $x^4 y^5 \cdot x^3 y^2$

   c) $x^3 \cdot x^5$

CP-16.   ♥ ♠ ♣ ♦ Suppose you have a standard deck of 52 playing cards.

   a)   What is the probability of randomly drawing a five of spades?

   b)   What is the probability of randomly drawing a five?

   c)   Suppose you draw a five and do not put the card back. What is the probability of randomly drawing another five?

   d)   Write each of the probabilities in parts (a), (b) and (c) as a percent.

CP-17.   Derek's scientific calculator displays 1.3 $^{-03}$. Explain what this means.

3.2 PRACTICE WRITING EQUATIONS AND THE DISTRIBUTIVE PROPERTY

CP-18.   **Algebra Tiles** For this problem, your instructor will provide you with a set of tiles in two shapes: rectangles and small squares. Suppose the small square has a side of length 1, and the rectangle has a side of unknown length, say x.

   a)   Use this information to find the area of each of the figures.

the small square: the rectangle:

□ 1   [   ] 1
   1   x

   b)   Trace each of the tiles in your Tool Kit. Mark the dimensions along each side, and clearly label each tile with its area. Algebra tiles are often referred to by their areas.

CP-19.   **Algebra Tiles and Area** Write an expression for the area of each of the following figures that are composed of Algebra Tiles. It may help you to build the figures with tiles first.

   a)   b)   c)   d)

CP-20.   **Algebra Tiles and the Distributive Property** Read the following example.

   Example: Note that "3(x)" means "3 times x" and can also be written 3x.

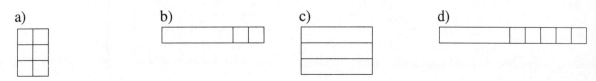

The area of this figure is three times the area of the figure in part (b) in the previous problem, so its area can be written as

3(area of b),
   or   3(x + 2).

The area of this figure is the sum of the areas in parts (a) and (c) in the previous problem, so its area can be written as

(area of a) + (area of c),
   or   3(x) + 3(2).

[PROBLEM CONTINUED ON NEXT PAGE]

CP-20. continued

> **Matching**: Now match each geometric figure on the left with an algebraic expression that describes it from the list on the right.

a) 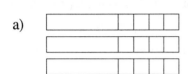   d)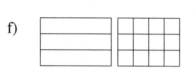

1. $2(x) + 2(2)$

2. $3(x + 4)$

3. $4(x) + 4(1)$

4. $3(x) + 3(1)$

b)    e)

5. $2(x + 5)$

6. $3(x) + 3(4)$

7. $5(x + 1)$

8. $3x + 9$

c)   f)

CP-21.   Sketch a geometric figure represented by each algebraic expression below.

a) $4(x) + 4(3)$ b) $4(x + 3)$

c)   What do you notice about the areas of the two figures in parts (a) and (b)? What can you conclude about the two expressions?

d) $3(x + 1)$ e) $3(x) + (3)(1)$

f)   What do you notice about the areas of the two figures in parts (d) and (e)? What can you conclude about the two expressions?

CP-22.   In the previous problem, you saw that

$$4(x) + 4(3) = 4(x + 3)\ \text{because they represent equal areas, and}$$

$$3(x + 1) = 3(x) + 3(1)\ \text{because they also represent equal areas.}$$

These are examples of the **Distributive Property** of multiplication over addition.

a)   Add the distributive property to your Tool Kit with examples and explanations to make it useful to you.

b)   Use the distributive property to rewrite each of the following expressions:

$7(3) + 7(x) = ??$

$5(x + 4) = ??$

CP-23.   Use the distributive property to match algebraic expressions in the left-hand column with equivalent expressions in the right-hand column. Not all expressions on the left can be matched with expressions on the right.

a.	$4(x + 5)$		A.	$x(x + 3)$
b.	$2(x + 3)$		B.	$-4(x) + (-4)(2)$
c.	$5(2x) + 5(4)$		C.	$(3 + x)3$
d.	$(x)(x) + (x)(3)$		D.	$4(x) + 4(5)$
e.	$2x(x + 4)$		E.	$5(x + 4)$
f.	$(3)(3) + (x)(3)$		F.	$2(x) + 2(3)$
g.	$-4(x + 2)$		G.	$(2x)(x) + (2x)(4)$

CP-24.   Use the distributive property to match algebraic expressions in the left-hand column with equivalent expressions in the right-hand column. Not all expressions on the left can be matched with expressions on the right. The expressions in the right column have been simplified.

a.	$4(x + 7)$		A.	$x^2 + y$
b.	$x(x + 1)$		B.	$x^2 + x$
c.	$(x + 9)2$		C.	$y^2 - 3y$
d.	$x(x + y)$		D.	$4x + 28$
e.	$3(2x + 5)$		E.	$10xy$
f.	$y(y - 3)$		F.	$6x + 15$
g.	$10(x + y)$		G.	$2x + 18$

CP-25.   You can illustrate the distributive property with Algebra Tiles and show, for example, that $6(x + 2) = 6x + 12$.

a)   Sketch two geometric figures which illustrate that $6(x + 2) = 6(x) + 6(2)$

You can also use the meaning of multiplication to show that $6(x + 2) = 6(x) + 6(2)$. For example, $3y$ means $y + y + y$.

b)   What does $6(x + 2)$ mean?

c)   Simplify the right-hand side of the equation you wrote in part (b) to show that $6(x + 2) = 6x + 12$.

CP-26.   In what way are these examples of the distributive property alike? How are they different?

$5(2 + 7) = 5(2) + 5(7)$

$5(x + 7) = 5(x) + 5(7)$

Solve each of problems CP-27 through CP-36 by making a Guess and Check Table and then write an equation from your table.

CP-27.   John has 380 out of 440 points to date in an algebra class. There are only 60 remaining points possible in the grading period. What is the least number of points John will need to have an average of 80% ?

CP-28.   A bank account has an effective annual yield of 5.15%. If the account has nearly $820.70 after one year, how much was initially invested in the account?

CP-29.   Larry and Moe leave Sacramento going in opposite directions. Larry drives five miles per hour faster than Moe. In four hours they are 476 miles apart. How fast is each person traveling?

CP-30.   Mr. Jordan keeps coins for paying the toll crossings on his commute to and from work. He presently has three more dimes than nickels and two fewer quarters than nickels. The total value is $5.40. Find the number of each type of coin.

CP-31.   Alejandra cut a 40-inch long board into two pieces and painted one piece purple and the other piece orange. The purple board is four inches longer than the orange board. How long is each painted board?

CP-32.   Raisa cut a string 112 centimeters long into two pieces so that one piece is three times as long as the other. How long is each piece?

CP-33.   Fred, the Chronicle newspaper distributor, collected all the dimes, nickels and quarters from one of his vending machines. He gathered twice as many dimes as quarters, and two more nickels than quarters. He collected a total of $7.60. How many coins did Fred take from the machine?

CP-34.   A fallen tree branch 94 inches long is cut into seven logs, all the same length. A 10" piece is left over. How long is each log?

CP-35.   The sum of the ages of Ruth and her mother is 77 years. The difference in their ages is 27 years. How old is each person?

CP-36.   Last year's pie-eating champion, Andy, trained seven consecutive days for this year's pie-eating contest. Each day he ate two more pies than the day before. Andy ate a total of 133 pies while in training. How many pies did he eat the first day of training?

CP-37.   **Algebra Tiles and Perimeter** The figures in parts (a) through (d) below are drawings of

rectangles built with algebra tiles: $\square$ 1 $\boxed{}$ 1   In problem CP-19, you found the area of each of the figures. This time, find the **perimeter** of each figure. It may help you to build the figures with tiles first.

a)

b)

c)

d)

CP-38.   Follow the instructions below for each of these equations:

   (1) $y = x$ (2) $y = x + 2$

   a)   Make a table with eight values for x between –3 and 3 and find the corresponding y-values.

   b)   Use your table to graph the equation. Your graph is incomplete without appropriate labels.

   c)   List two similarities and one difference you see in the graphs.

CP-39.   Find the value of each expression below for $x = 2$ and for $x = -3$.
          Note: You will compute two different values for each expression.

   a) $x^2 - 3x + 8$

   b) $-3x^2 + x$

   c) $x^2 + x - 6$

CP-40.   Put your calculator in scientific mode. For each problem below, copy the problem, use your calculator to compute the result, and write the result on your paper.

   a) $(2.1 \cdot 10^5)(3.25 \cdot 10^5) =$

   b) $(2.1 \cdot 10^5) + (3.25 \cdot 10^5) =$

   c)   Write the results for parts (a) and (b) in standard form.

3.3 SOLVING EQUATIONS USING CUPS AND TILES

CP-41.   Build each equation with cups and tiles, then solve it as we did in class. Be sure to record each of the moves you make in solving an equation by making a sketch **and** writing what you did at each step. Show your check!

a) $2x + 3 = 11$

b) $2x + 1 = 13$

c) $5x + 8 = 3x + 10$

d) $5 + 2x = 7 + x$

CP-42.   Use the idea of cups and tiles to solve each of the following equations.

a) $2x + 1 = 12$

b) $-3 + 2b = -9$

c) $-5 = 3m + 1$

d) $-6 + 4y = 3$

e) $c - 8 = 2 + 5c$

CP-43.   Solve each of the following equations by "inspection" (mental math).

a) $x + 7 = 2$

b) $3 + x = -4$

c) $5 = x - 1$

d) $4 = -2 + x$

e) $2 = x + 5$

CP-44.   Solve each of the following equations by any method you choose.

a) $-15 = d - 24$

b) $-13 = k - 18$

c) $p - 12 = -15$

d) $-9 + p = -14$

e) $-6.2 = q - 4.4$

> For each of problems CP-45 and CP-46,
>
>   a) use a Guess and Check Table to write an equation;
>
>   b) use algebra to solve the equation;
>
>   c) write your answer carefully; and
>
>   d) check your answer.

CP-45.   The length of a rectangle is three times the width. The perimeter is 16 feet. What is the width of the rectangle?

CP-46.   One number is three more than twice another number. The sum of the numbers is 39. What are the numbers?

CP-47.   Express each of the following products in a simpler exponential form.

    a) $(2^3)(2^4)$ d) $(x^2)^5$

    b) $(x^3)(x^4)$ e) $\dfrac{2x^4}{2x^2}$

    c) $(2^3)^4$ f) $\dfrac{x^3y^4}{xy^2}$

CP-48.   Make each of the equations below true by inserting one or more sets of parentheses.

    a) $2 \cdot 3^2 + 4 \cdot 3 - 1 = 47$

    b) $2 \cdot 3^2 + 4 \cdot 3 - 1 = 80$

CP-49.   Follow the instructions below for each of these equations:

    (1) $y = 2x$ (2) $y = x^2$

    a)   Make a table with eight values for x between −3 and 3 and find the corresponding y-values.

    b)   Use your table to graph the equation. Your graph is incomplete without the appropriate labels.

    c)   What makes one of the graphs a straight line and the other a curve (in this case a parabola) ?

CP-50.   The two line segments below have equal lengths.

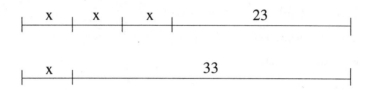

Write an equation that represents the diagram. How long is x ?

CP-51.   Hakeem has ten coins in his pocket, all quarters, nickels, and dimes, worth a total of $1.10. When he reaches in his pocket to grab one coin, the probability that he grabs a nickel is the same as the probability that he grabs a dime. What is the probability that he will grab a quarter?

CP-52.   Use your pattern-detection skills to write an equation represented by the Guess and Check Table below. Then write a word problem that fits. (There are many possible word problems.)

Guess First number	Second number	Total	Check = 149 ?
65	68	133	too low
71	74	145	too low
75	78	153	too high
73	76	149	just right

Answer: The numbers are 73 and 76.
Equation:
Your word problem:

CP-53.   What is the ratio of five days to two weeks (in days)?

CP-54.   What is the ratio of four inches to one foot (in inches)?

CP-55.   Jack Hammer does a job in 25 minutes. Peter Piper does the same job in half an hour. Write a ratio to …

a)   compare Hammer's time to Piper's time.

b)   compare Piper's time to Hammer's time.

3.4 SOLVING COMPLICATED EQUATIONS

CP-56.   Solve each of the following equations for x.

    a) $3x + 2 = 14$

    b) $-2x + 6 = 10$

    c) $-2x - 4 = 4x + 8$

    d) $3x - 1 = 5x + 6$

CP-57.   For each equation below, first describe in words what the equation says, and then describe what you will do to solve it. Finally, solve each equation for the given variable.

    a) $-3x = 13$

    b) $\frac{1}{2}y = -13$

    c) $\frac{2}{3}z = 8$

    d) $\frac{2}{3}z - 2 = 8$

    e) $\frac{7}{2}x + 6 = x + 12$

CP-58.   Solve each of the following equations for x.

    a) $x + \frac{2}{3} = 1\frac{1}{2}$

    b) $5x + 0.5 = 0.2x + 1.5$

    c) $\frac{2}{5}x = 5$

    d) $0.3x + 1.2 = x + 8.4$

    e) $x + 5 = 1$

CP-59.   Solve each equation below by any method you choose.

    a) $x + 8 = 2$

    b) $c - 1 = -4 - 2c$

    c) $-9.8 = p - 5.5$

    d) $y - 5\frac{3}{4} = -3\frac{1}{4}$

    e) $x - 8\frac{1}{2} = -5\frac{1}{6}$

For each of problems CP-60 through CP-63, write an equation and solve it. Be sure to identify what the variable represents if you do not make a Guess and Check Table.

CP-60.   Find three consecutive even numbers whose sum is 54.

CP-61.   The length of a rectangle is twice its width. Its perimeter is 36 centimeters. Find the area of the rectangle.

CP-62.   A large cake is cut into four pieces so that each piece is twice as heavy as the preceding one. The entire cake weighs five pounds. How many ounces does each piece weigh? (Hint: The problem will be easier if you choose your variable x carefully.)

CP-63.   A stick 152 centimeters long is cut up into four short pieces, all the same length, and two longer pieces (both the same length). A long piece is 10 centimeters longer than a short piece. Into what length pieces is the stick cut?

CP-64.   Use the idea of cups and tiles to solve each of the following equations.

   a) $4x - 3 = 5 + 2x$

   b) $4 - 2x = 4x - 8$

   c) $3 + 3x = 9 + 2x$

   d) $7 - 2x = 3 + 3x$

   e) $3x - 3 = 6 - 3x$

CP-65.   The two line segments below have equal lengths.

   a)   Find the length x.

   b)   Write a sentence to compare how you solved part (a) with what you did in problem CP-64(c).

CP-66.   Use the distributive property to rewrite each expression.

   a) $3(x + 4)$

   b) $b(a - 7)$

   c) $x(x - 3)$

   d) $-7(c + 8)$

   e)   Reverse the process: $4x + 8 = 4(\_\_ + \_\_)$

   f) $5x^2 + 10x = 5x(\_\_ + \_\_)$

   g) $3x + 9 = \_\_(\_\_ + \_\_)$

CP-67.   Use your pattern-detection skills to write an equation and a word problem represented by the Guess and Check Table below. There are many word problems you could write. Write only one, but be creative!

Guess First number	Second number	Product	Check = 1974 ?
30	35	1050	too low
40	45	1800	too low
43	48	2064	too high
42	47	1974	correct

   Answer: The number is 42.
   Equation:
   Your word problem:

CP-68.   Alberto collects information about the speed of a roller coaster compared to the location of the roller coaster along its track. Can Alberto use his data to draw a continuous graph, or should it be discrete? Explain your answer.

   For each of problems CP-69 and CP-70, write an equation and solve it. Be sure to define the variables you use.

CP-69.   A certain rectangle has a perimeter of 26 centimeters. Its width is 7.25 centimeters. Draw a diagram. Find the length and the area of the rectangle.

CP-70.   The coins in a piggy bank are all dimes, nickels, and quarters. There are twice as many dimes as quarters, and two more nickels than quarters. If the total value of the coins is $7.60, how many coins are in the piggy bank?

CP-71.   Graph each of the following equations on the same set of axes. Remember to label the axes and each graph.

a) $y = x$

b) $y = x - 3$

c) $y = 3x$

d)   Compare the graphs in parts (a) and (b): How did the "3" in part (b) change the graph from part (a)?

e)   Compare the graphs in parts (a) and (c): How did the "3" in part (c) change the graph from part (a)?

CP-72.   Use your pattern-detection skills to write an equation and a word problem represented by the Guess and Check Table below. There are many word problems you could write.

Guess First number	Second number	Total	Check = 22 ?
4	9	13	too low
5	11	16	too low
9	19	28	too high
7	15	22	correct

Answers: The two numbers are 7 and 15.
Equation:
Your word problem:

3.5 SOLVING EQUATIONS CONTAINING PARENTHESES

CP-73.   Copy this example in your notebook:

Solving Equations containing Parentheses

Example: Solve the equation $5(2x + 7) = x + 71$. $5(2x + 7) = x + 71$

The feature that makes this equation different from the ones you've solved before are the parentheses. You must deal with them first.

In order to get rid of the parentheses, use the distributive property to
rewrite the left side of the equation: $10x + 35 = x + 71$

Now the equation looks like many you have seen before and you can
solve it in the usual way:
subtract 35 from each side and subtract x from each side, and $9x = 36$
then divide both sides of the resulting equation by 9. $x = 4$

CP-74.   You could mimic the process used in CP-73 to solve the equation $9(3x - 8) = 36$:

Copy the table shown below on your paper.
Fill in the blank labeled (b) to explain how the equation to its right was obtained from the
equation above it. Then do the same for (c).

		$9(3x - 8) = 36$
a)	Distribute the 9.	$27x - 72 = 36$
b)	_____	$27x = 108$
c)	_____	$x = 4$

CP-75.   Here's another way to solve the equation $9(3x - 8) = 36$:

Copy the table shown below on your paper.
Fill in the blank labeled (a) to explain how the equation to its right was obtained from the
equation above it. Then do the same for (b), and finally for (c).

		$9(3x - 8) = 36$
a)	_____	$3x - 8 = 4$
b)	_____	$3x = 12$
c)	_____	$x = 4$

CP-76.   a)   Describe how the equation solving methods in CP-74 and CP-75 are different. In what ways are the two methods alike?

b)   Explain why the method shown in problem CP-75 might be inconvenient for solving the equation $3(2x - 1) = 7$.

c)   Explain why the method shown in problem CP-74 might be inconvenient for solving the equation $1250(657x - 1324) = 1250$.

CP-77.   Solve each of the following equations for x. Show each step you use.

a) $6(5x + 12) = 162$

b) $2(7x + 15) = x + 121$

c) $5(6x - 9) = -15 + 2x$

d) $4(9x - 14) = 16$

CP-78.   Solve each of the following equations for x. Show the steps you use.

a) $5(x - 4) = x + 25$ d) $3(4x + 1) = 163$

b) $7(2x - 4) + 3 = 31$ e) $6(3x + 2) - 18 = 7$

c) $3(7x + 9) = -15$ f) $4 - 8(3x - 5) = 92$

Hint for part (f): Remember that $4 - 8(3x - 5)$ means $4 + (-8)(3x - 5)$, and the multiplication is done **first**.

CP-79.   Copy the following examples in your notebook. Fill in each blank to explain how the equation on its right was formed.

Example 1: Solve the equation $2(x + 4) = 13$.

$$2(x + 4) = 13$$

i)   Distribute the 2. $2x + 8 = 13$

ii)   _____ $2x = 5$

iii)   _____ $x = \dfrac{5}{2}$

[PROBLEM CONTINUED ON NEXT PAGE]

CP-79. continued

Example 2: Solve the equation $2(x + w) = 13$ for x.

$$2(x + w) = 13$$

i)   Distribute the 2. $2x + 2w = 13$

ii)   _____ $2x = 13 - 2w$

iii) _____ $x = \dfrac{13 - 2w}{2}$

Now, solve each of the following equations for x.

a) $2x + 3 = 13$ e) $5x + 4 = x + 20$

b) $2x + b = 13$ f) $5x + d = x + 20$

c) $3x - 2 = 10$
                                            g) $\frac{x}{2} + 5 = 9$

d) $bx - 2 = 10$
                                            h) $\frac{x}{2} + e = 9$

CP-80.   Look back at problem CP-79. Compare how you solved the equation in part (b) with how you solved the equation in part (a). Write one or two sentences to explain how the processes you used to solve each pair of equations were alike and how they were different.

CP-81.   a)   Solve $3x + 91 = 43$ for x.

          b)   Solve $3x + c = 43$ for x.

          c)   Solve $3x + c = 43$ for c.

          d)   Solve $5x + 3c = 17$ for x, then solve it for c.

CP-82.   For each equation below, make a table with entries for values of x and compute the corresponding values for y. Graph each equation on a separate set of axes. Be sure everything is labeled.

          a) $y = x^2 - 3x + 2$ for $-1 \le x \le 4$. Explain why it is important to use 1.5 as an input value.

          b) $y = \frac{1}{3}x + 3$ (Be sure to use negative as well as positive numbers.)

          c) $y = 8 - x$

CP-83.   **Area Computation** The large rectangle shown below has been fractured into smaller rectangular parts. Here's an example of how we can compute the area of the whole rectangle in two ways:

Method 1: We can find the total area by first finding the area of each part and then **adding**:

$$100 + 30 + 40 + 12 = 182$$

Method 2: We can find the total area by **multiplying** the length of the whole rectangle by the width of the whole rectangle:

$$(10 + 3)(10 + 4) = (13)(14) = 182.$$

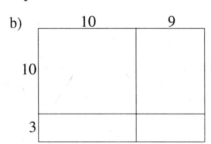

Draw each of the following fractured rectangles on your paper and find the area of each figure in two ways:

(1)   by adding the areas of the parts, and

(2)   by multiplying the length of the whole rectangle by the width of the whole rectangle.

Show your work as in the example.

a)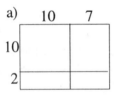

b)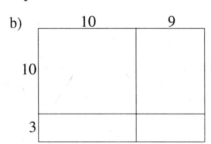

c)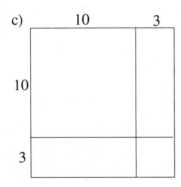

d)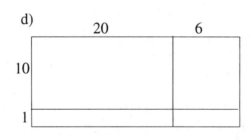

CP-84.   Explain why Figures A and B have the same area. Does your reasoning also apply to Figures C and D ? Explain.

100	50
60	30

Figure A

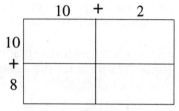

Figure B

100	20
80	16

Figure C

Figure D

For each of problems CP-85 through CP-87, write an equation and solve it. Be sure to define the variables you use.

CP-85.   If one side of a square is increased by 12 feet and an adjacent side is decreased by three feet, a rectangle is formed whose perimeter is 64 feet. Draw a diagram and find the length of the side of the original square.

CP-86.   In honor of the 27th anniversary of Malwart Department Store, everything was 27% off. Jarlene bought a jacket at the sale for $45.26. What was the original price of the jacket?

CP-87.   Moe started washing dishes at Larry's Café. He washed nine dishes per minute. Fifteen minutes later Curley joined Moe and washed 16 dishes per minute. They washed a total of 760 dishes before stopping to rest. How long did Curley work? How many dishes did Moe wash?

CP-88.   Pat found a die which had been made badly: it had two 6's on it and no 1's. If Pat rolls the die, how likely is it to come up …

a)   a 2 ?

b)   a 6 ?

c)   either a 5 or a 6 ?

3.6 CHOOSING A PHONE PLAN

CP-89.   **Choosing A Phone Plan** In a college dormitory, each student has a choice of two phone companies. Company A charges $7.46 per month plus 13 cents a call, while Company B charges $6.17 per month plus 17 cents per call.

   a)   About how many phone calls do you make per month?

   b)   For each company, write an equation which represents the cost in a given month in terms of the number of phone calls.

   c)   Graph each equation you wrote in part (b). Label the vertical axis "cost per month" and the horizontal axis "number of calls" per month.   Be sure to label each graph.

   d)   Discuss how your two graphs relate to the solution of the problem. When are the costs for both companies the same? When is Company A a better choice? When is Company B a better choice?

   e)   How many calls do you think that the average student in your class makes?

   f)   How could you find out the answer to (e)?

   g)   Carry out your plan from part (f).

   h)   Decide which students in your class should use which phone company and tell why.

CP-90.   Solve each of the following equations.

   a) $3(x + 1) = 4$

   b) $4(y - 2) + 3 = 19$

   c) $7(w + 5) - 2 = 6$

   d) $-2(x - 2) = 16$

   e) $3(2x + 4) = 28$

CP-91.   Solve each of the following equations by any methods you choose.

   a) $\frac{1}{2}x + 4 = -\frac{1}{4}x + 7\frac{1}{2}$

   b) $x + .63 = 1.56$

   c) $0.78x - 2 = 0.8x + 8.4$

   d) $0.38x = 1.82$

   e) $1.2 - x = 0.8x - 1.2$

   f) $3x = 2$

   g) $\frac{x}{1.5} = 2$

> For each of problems CP-92 through CP-94, write an equation and solve it. Be sure to define the variables you use.

CP-92.   Each side of a square garden is increased by three meters. The perimeter of the new garden is 50 meters. What was the length of the original square garden? Draw a diagram and label the sides.

CP-93.   Mrs. Agnos keeps cats and canaries. The animals have a combined total of 30 heads and 80 legs. How many cats does Mrs. Agnos have?

CP-94.   Sunrise Sand and Gravel charges $4.50 for each cubic yard of sand plus $6.00 to deliver the sand. How much sand could you have delivered for $100.00?

CP-95.   Rewrite each of the expressions below in a simpler exponential form, if possible. Otherwise, write an expression in standard form.

   a) $3^4 \cdot 3^5$ d) $3^4 + 3^5$

   b) $(x^3)^4$ e) $(10^3)^5$

   c) $\dfrac{6x^3y^2}{3x^2y^2}$ f) $10^5 - 10^3$

CP-96.   Use any method you like to solve each of the following equations.

   a) $3x + 9 = 11$

   b) $8 = 4(x + 7) - 20$

   c) $\dfrac{3x}{4} = 5$

   d) $\dfrac{x - 3}{2} = 3$

CP-97.   Use any method you like to solve each of the following equations.

   a) $3x = 9$

   b) $5x = 11$

   c) $14 = -3x$

   d) $\dfrac{2}{5}x = 6$

   e) $\dfrac{9}{14} = \dfrac{-3}{7}x$

CP-98.   Use any method you like to solve each of the following equations.

   a) $5(4x + 3) = 75$

   b) $-6(3x - 8) = -6$

   c) $3 + 4(x + 1) = 159$

   d) $-3(8x - 4) = 18$

   e) $2 - 3(2x - 1) = 17$ (Hint: Look back at what you did in problem CP-78(f).)

CP-99.   **Chapter 3 Summary: Rough Draft** The three main ideas of Chapter 3 are:

   1.   Writing equations from Guess and Check Tables;

   2.   Using the distributive property to rewrite expressions; and

   3.   Solving linear equations.

Write your answers to the following questions in rough draft form **on separate sheets of paper**, and be ready to discuss them with your group at the next class meeting. Focus on the **content**, not neatness or appearance, as you write your summary. You will have the chance to revise your work after discussing the rough draft with your group.

   a)   For each of these three ideas, select and neatly recopy a problem that you did well, and write why you think that the selected problem is a good representative of the idea.

   b)   In this chapter you have looked at solving linear equations. Write a paragraph to explain how solving a linear equation is like unwrapping a present. You may wish to include an example to illustrate your ideas.

   c)   What were the most difficult parts of this chapter? List sample problems and discuss the hard parts.

   d)   What problem did you like best and what did you like about it?

3.7 SUMMARY AND REVIEW

CP-100. **Chapter 3 Summary: Group Discussion** Take out the rough draft summary you completed in CP-99. Use some class time to discuss your work; use homework time to revise your summaries as needed.

For each of the three main ideas of the chapter, choose one member of the group to lead a short discussion. The discussion leaders should take turns to:

- explain the problem they chose to illustrate their main idea,

- explain why they chose that particular problem,

- tell which problem they liked best and what they liked about it, and

- tell what they thought were the most difficult parts of this chapter.

This is your chance to make sure your summary is complete, update your Tool Kits, and work together on problems you may not be able to solve yet.

CP-101. Solve each of the equations below for x. Leave your answers in fraction form.

For example, if $2x + 3 = 0$, then $x = -\frac{3}{2}$.

a) $2x + 5 = 0$ e) $3x - 5 = 0$

b) $3x + 1 = 0$ f) $6x + 1 = 0$

c) $4x - 1 = 0$ g) $3x + 2 = 0$

d) $5x + 2 = 0$ h) $2x - 7 = 0$

CP-102. a) Describe a pattern that relates each equation in problem CP-101 to its solution.

b) Use the pattern you described in part (a) to solve $9667x - 555 = 0$.

c) Use your pattern to solve $ax + b = 0$ for x.

CP-103. Each of the equations below has two solutions. Use a Guess and Check Table to find the solutions.
For example, for $(x + 1)(x - 2) = 0$, we could use this table:

Guess x	x + 1	x - 2	Product	Check = 0 ?

a) $(x + 4)(x - 5) = 0$

b) $(x - 2)(x + 3) = 0$

> For each of problems CP-104 and CP-105, write an equation and solve it. Be sure to define the variables you use.

CP-104.   On an algebra test each question in Part A is worth three points and each question in Part B is worth five points. Mabel answered 19 questions correctly and had a score of 81. How many questions on each part of the test did she answer correctly?

CP-105.   Linda and Paul are recording artists. Last month their combined income was $13,000. Linda made $1,000 more than five times what Paul made. How much money did Paul earn?

> Solve each of problems CP-106 through CP-109 and write an equation for each one.

CP-106.   Choco-nut cookies cost $0.20 each plus $0.15 for the box . If one box of cookies costs $3.55, how many cookies do you get in a box?

CP-107.   A 112-centimeter board is cut so one piece is three times as long as the other. How long is each piece?

CP-108.   The perimeter of a triangle is 33 centimeters. The second side is twice as long as the first side, and the third side is eight centimeters longer than the second. How long is each side?

CP-109.   Orlando's savings account has a yield of 8%. After one year of not adding or withdrawing from it the balance is $1350.00. What amount was in the account at the start of the year?

CP-110.   Solve each of the following equations for x.

   a) $3x = b$

   b) $2x + 7 = p$

   c) $3x - 2 = x + q$

   d) $\frac{x}{2} + 5 = 2r$

   e) $\frac{2}{3}x - 6 = 4s$

CP-111.   Solve each equation for x. Show your steps.

   a) $\frac{2}{3}x = 9$ d) $3(x + 4) = 18$

   b) $2x - 3 = 5$ e) $2 = 2(x+3)$

   c) $5x + 8 = 1.7$

CP-112.   Solve each equation for x. Show your steps.

    a) $3 - 2(x - 5) = 15$ d) $8 = -4 + 3(3x + 5)$

    b) $0 = 2x + 3$ e) $3(3x + 5) - 4 = 8$

    c) $2(x + 3) - 5 = 0$

CP-113.   Copy and solve these Diamond Problems:

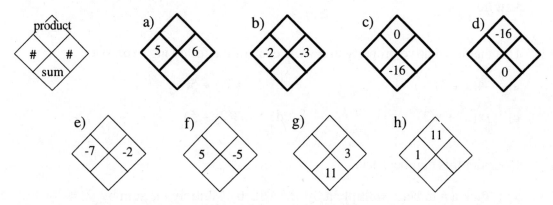

CP-114.*   Examine each graph and then write a story or description about what each graph shows. Your
           story may be different from those of others in your group.

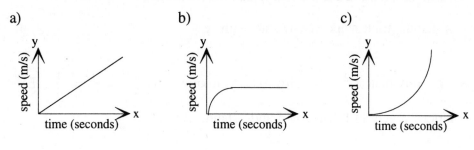

CP-115.   Solve each of the following equations for x.

    a) $x + 1.5 = -3.25$ b) $4x + 5 = 2x + 7$

    c) $8 = 4(x + 7) - 20$ d) $3x + c = 10$

    e) $2x + 3 = 10$ f) $15 = 6x + 4x - 8$

    g) $\dfrac{3x + 5}{4} = 5$ h) $xc - 5 = y$

    i) $\dfrac{3}{4}x - 6 = 12$ j) $14 - 3(2x + 1) = 15$

    k) $4x + c = -x + 7$

---

* Adapted from *The Language of Functions and Graphs*, Joint Matriculation Board and the Shell Centre for Mathematical
  Education, University of Nottingham, England

CP-116.   Write each of the following expressions as a power, if possible.

    a) $x^2 \cdot x^5$

    b) $\dfrac{10^6}{10^{-2}}$

    c) $(x^3)^4$

    d) $x^2 + x^3$

CP-117.   Sketch a graph of the relationship between your test scores and the amount of homework that you do.

CP-118.   Use the distributive property to rewrite each of the following expressions.

    a) $4(x + 5)$

    b) $5(2x) + 5(4)$

    c) $x(x) + x(3)$

    d) $-3(2y - 5)$

CP-119.   Consider this number trick:

> "Pick a number. Multiply it by 2. Add 6. Multiply the sum by 2.
> Divide by 4, then subtract 3. The result equals your starting number."

    a)   Choose a number and use it to show that the trick works.

    b)   Show that the trick works if you start with x.

CP-120.   Some people think that $2^2 = -4$. However, you know that $2^{-2} = \dfrac{1}{2^2} = \dfrac{1}{4}$. Explain, in a way that another algebra student could understand, why $2^{-2} = \dfrac{1}{4}$.

CP-121.   **Chapter 3 Summary: Revision** This is the final summary problem for Chapter 3. Using your rough draft from CP-99 and the ideas you discussed in your groups from CP-100, spend time revising and refining your Chapter 3 Summary. Your presentation should be thorough and organized, and should be done on a separate piece of paper.

Algebra Tiles

Glue this page to a piece of lightweight cardboard (an empty cereal box works) or another piece of paper. Carefully cut out the individual rectangles and the small and large squares. Keep them in a re-sealable envelope or plastic bag. Store the bag of tiles in your notebook.

Algebra Tool Kit: the "what to do when you don't remember what to do" kit

Algebra Tool Kit: the "what to do when you don't remember what to do" kit

Calculator Tool Kit: for instructions on using your calculator

Chapter 4

ESTIMATING FISH POPULATIONS: Numerical, Geometric, and Algebraic Ratios

CHAPTER 4

ESTIMATING FISH POPULATIONS:
NUMERICAL, GEOMETRIC, AND ALGEBRAIC RATIOS

In this chapter you will be exploring the
concept of **ratio**, an idea which is useful
in many aspects of our lives. You'll use
skills you develop to solve a problem
similar to this:

> Fish biologists need to keep track of fish populations in the waters they
> monitor. They might want to know, for example, how many striped bass
> there are in San Francisco Bay. This number changes, however,
> throughout the year, as fish move in and out of the bay to spawn.
> Therefore biologists need a way to gather current data fairly quickly and
> inexpensively.

One of the goals in this chapter is to develop skills to solve this problem efficiently.

In this chapter you will have the opportunity to:

* enlarge and reduce geometric shapes and see how the ratios of corresponding sides,
  perimeters, and areas are related,

* study similar triangles and explore the idea that the ratios of corresponding sides of
  similar triangles are equal,

* represent ratio problems by the graph of a line passing through the origin, and thus
  connect ratios with graphing ideas from Chapter 3,

* use equivalent ratios to write equations to solve problems, and

* view percent problems as ratio problems.

CHAPTER CONTENTS

4.1 ENLARGING AND REDUCING GEOMETRIC FIGURES

EF-1.   How do you think a fish biologist might estimate the number of fish in the San Francisco Bay?

EF-2.   **Enlarging Geometric Figures** Architects or graphic artists sometimes need to make a small-scale drawing from a larger one, or a larger drawing from a small blueprint or plan. In this section, you will see how geometric ratios can be used to make **enlargements** or **reductions** of geometric figures.

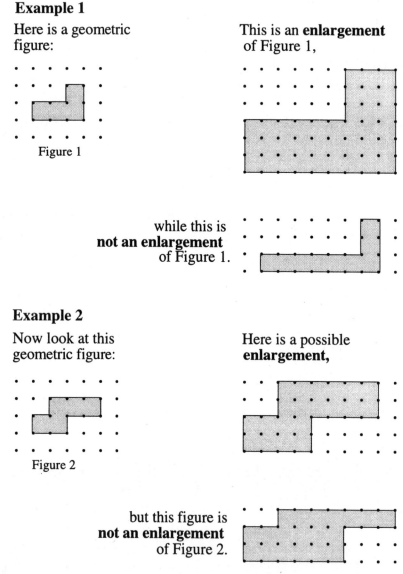

Example 1

Here is a geometric figure:

Figure 1

This is an **enlargement** of Figure 1,

while this is **not an enlargement** of Figure 1.

Example 2

Now look at this geometric figure:

Figure 2

Here is a possible **enlargement,**

but this figure is **not an enlargement** of Figure 2.

Write one or two sentences to explain why the third figure in Example 2 is not an enlargement of Figure 2.

EF-3.   **Enlargement Ratios** Lissa saw a dot grid figure and decided to enlarge it. She made the corresponding sides of the new figure **twice** as long as those in the original figure. In other words, the enlargement ratio for the sides was $\frac{2}{1}$. Here are Lissa's results:

original figure

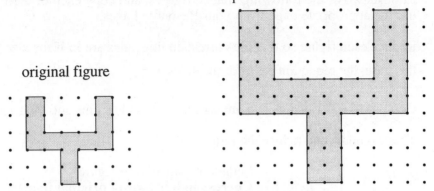

Select two of the following figures. Then follow these directions for each figure you selected:

1)   Copy the original figure on dot paper, and then enlarge the figure by making the corresponding sides of the new figure **twice** as long as those in the original figure. In other words, use a side enlargement ratio of $\frac{2}{1}$ to enlarge your copy of the original figure.

2)   Compute the perimeters and areas of both the original figure and its enlargement.

3)   Finally, for each figure, find and simplify the following ratios:

$$\frac{\text{Length of \textbf{Side} of \textbf{new} figure}}{\text{Length of \textbf{Corresponding Side} of \textbf{original} figure}},$$

$$\frac{\text{\textbf{Perimeter} of \textbf{new} figure}}{\text{\textbf{Perimeter} of \textbf{original} figure}}, \quad \text{and} \quad \frac{\text{\textbf{Area} of \textbf{new} figure}}{\text{\textbf{Area} of \textbf{original} figure}}.$$

a)   b)   c)   d)

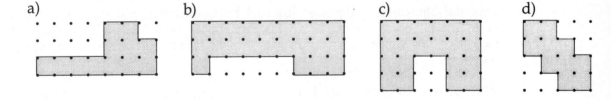

EF-4.   **Reducing Figures** Follow the directions listed below to apply what you know about enlarging figures to making copies that are smaller:

1)   Choose two of the following shaded figures, and copy each of them on dot paper. Or, use the appropriate page from your Resource Pages.

2)   Reduce each figure so that the corresponding sides are **half** the size of the original. In this case, the reduction ratio for the sides is $\frac{1}{2}$.

3)   Compute the perimeters and areas of the original figures and their reductions.

4)   Then compute and reduce the ratios:

$$\frac{\text{Length of \textbf{Side} of \textbf{new} figure}}{\text{Length of \textbf{Corresponding Side} of \textbf{original} figure}},$$

$$\frac{\textbf{Perimeter of new} \text{ figure}}{\textbf{Perimeter of original} \text{ figure}}, \text{ and } \frac{\textbf{Area of new} \text{ figure}}{\textbf{Area of original} \text{ figure}}.$$

a)   b)   c)

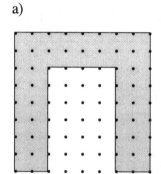   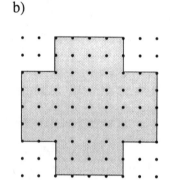   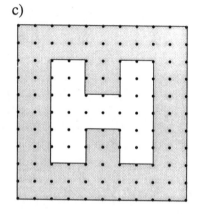

EF-5.   **Percents**

a)   Max, the inquisitive third grader, just took his first spelling test and got 80 out of a possible 100 points. When he got his paper back, "80%" was written at the top of the page and Max wondered what it meant. Explain it to him in one or two complete sentences.

a)   While reading the newspaper, Laci noticed a whole-page ad for Malwart's that declared all merchandise was half price. At the bottom of the page, the ad concluded, "Hurry! Every item 50% its usual price!" Laci wondered why "half price" and "50% price" meant the same thing. Explain to her in one or two complete sentences why $\frac{1}{2} = 50\%$.

EF-6.   Solve each of the following equations for x.

a) $\dfrac{x}{100} = \dfrac{43}{100}$ b) $\dfrac{x}{100} = \dfrac{7}{100}$

c) $\dfrac{x}{100} = \dfrac{6}{10}$ d) $\dfrac{x}{100} = \dfrac{3}{25}$

e) $\dfrac{x}{100} = \dfrac{3}{5}$ f) $\dfrac{x}{100} = \dfrac{9}{20}$

g) $\dfrac{x}{100} = \dfrac{5}{8}$ h) $\dfrac{x}{100} = \dfrac{2}{3}$

i) $\dfrac{x}{100} = \dfrac{9}{4}$ j) $\dfrac{x}{100} = \dfrac{7}{5}$

EF-7.   Suppose a 4' × 5' rectangle is enlarged so that the ratio of corresponding sides is $\dfrac{4}{1}$.

a)   What are the dimensions of the new rectangle?

b)   Compute the perimeter and the area of original rectangle and the new, enlarged rectangle.

c)   Compute the following ratios:

$$\frac{\text{Length of \textbf{Side} of \textbf{new} figure}}{\text{Length of \textbf{Corresponding Side} of \textbf{original} figure}},$$

$$\frac{\text{\textbf{Perimeter} of \textbf{new} figure}}{\text{\textbf{Perimeter} of \textbf{original} figure}}, \quad \text{and} \quad \frac{\text{\textbf{Area} of \textbf{new} figure}}{\text{\textbf{Area} of \textbf{original} figure}}.$$

EF-8.   Derek took two minutes to do his homework. Dean took five seconds. Find a ratio of Derek's time to Dean's time in terms of seconds. Be careful.

EF-9.   Solve the following problem and write an equation for it.

A bucket filled with water weighs 8.4 kilograms. If the water by itself weighs five times as much as the bucket, what is the weight of the bucket?

EF-10.   Express each of the following ratios as a percent. For example, $\dfrac{1}{5} = \dfrac{20}{100} = 20\%$.

a) $\dfrac{1}{4}$ c) $\dfrac{17}{25}$

b) $\dfrac{3}{10}$ d) $\dfrac{9}{5}$

EF-11.   **Subproblems** A useful strategy for solving problems is to identify and solve "smaller" or "simpler" related subproblems. Here is one way this strategy can be applied:

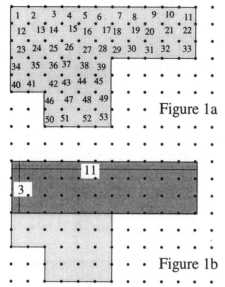

Figure 1a

Figure 1b

Max, who is in the sixth grade, found the shaded area in Figure 1a to be 53 square units by counting each individual square.

Max's older sister, Maxine, claimed that she could find the area more easily -- by dividing the original problem into **subproblems**.

The first subproblem she solved was to find the area of the rectangle at the top of Figure 1b:

She reasoned that the length is 11 and the width is 3, so the area of that rectangle is $3 \cdot 11 = 33$ square units.

Maxine is certain you can finish the problem in just two "steps" -- two subproblems, that is.

Identify two subproblems you could use to finish computing the area of Figure 1b, solve each of them, and then check Max's work by finding the total shaded area.

EF-12.   Use the idea of subproblems to find the total shaded area in Figure 2 and in Figure 3. Show all of your work by stating each subproblem.

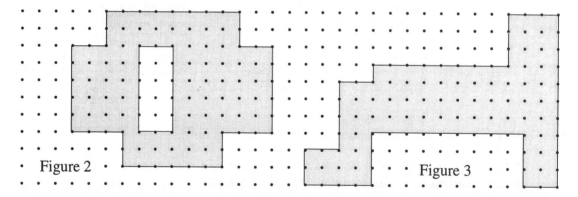

Figure 2

Figure 3

EF-13.   Rewrite each of the following products.

a) $10(5 + 3)$

b) $10(n + 3)$

c) $4(x + 1)$

d) $3(y + 5)$

e) $5(y + 3)$

EF-14.   Sketch a geometric figure described by each of the following expressions, and then find each product using the area of each figure.

   a) $10(10 + 3)$

   b) $2(10 + 3)$

   c) $(10 + 2)(10 + 3)$

EF-15.   Solve each of the following equations for x.

   a) $4(2x + 7) = 108$

   b) $9(19x - 4) = 3x$

   c) $15 - (2x + 7) = 14$ (Hint: Why does $-(2x + 7) = -2x - 7$?)

   d) $14x - 53 - 32x = 73$

EF-16.   a)   Graph the curve $y = x^2 - 3$. Choose values for x so that $-3 \le x \le 3$.

   b)   Estimate the points where the parabola in part (a) crosses the x-axis.

4.2 MORE ENLARGING, REDUCING, AND RATIOS

EF-17.   Look back at EF-12 and discuss the following questions with your group. How many different ways of finding the shaded areas did your group find? How did you divide up the larger problem into subproblems? Was one way more efficient than the other ways? Be ready to share your methods with the rest of the class.

EF-18.   a)   Draw a rectangle with a perimeter of 24 units and an area of 32 square units on dot paper.

b)   Reduce the rectangle you drew in part (a) so that the ratio of corresponding sides is $\frac{1}{2}$. Then compute the ratios:

$$\frac{\text{Length of } \textbf{Side} \text{ of } \textbf{new} \text{ figure}}{\text{Length of } \textbf{Corresponding Side} \text{ of original figure}},$$

$$\frac{\textbf{Perimeter} \text{ of } \textbf{new} \text{ figure}}{\textbf{Perimeter} \text{ of original figure}}, \text{ and } \frac{\textbf{Area} \text{ of } \textbf{new} \text{ figure}}{\textbf{Area} \text{ of original figure}}.$$

c)   Reduce the rectangle you drew in part (a) so that the ratio of corresponding sides is $\frac{1}{4}$. Then compute the ratios

$$\frac{\text{Length of } \textbf{Side} \text{ of } \textbf{new} \text{ figure}}{\text{Length of } \textbf{Corresponding Side} \text{ of original figure}},$$

$$\frac{\textbf{Perimeter} \text{ of } \textbf{new} \text{ figure}}{\textbf{Perimeter} \text{ of original figure}}, \text{ and } \frac{\textbf{Area} \text{ of } \textbf{new} \text{ figure}}{\textbf{Area} \text{ of original figure}}.$$

EF-19.   **Notation** To save time and effort, instead of writing

$$\frac{\text{Length of } \textbf{Side} \text{ of } \textbf{new} \text{ figure}}{\text{Length of } \textbf{Corresponding Side} \text{ of original figure}}$$

we will agree to use the notation $\dfrac{S_{new}}{S_{original}}$.

Similarly, instead of $\dfrac{\textbf{Perimeter} \text{ of } \textbf{new} \text{ figure}}{\textbf{Perimeter} \text{ of original figure}}$, we will agree to use the notation $\dfrac{P_{new}}{P_{original}}$,

and for $\dfrac{\textbf{Area} \text{ of } \textbf{new} \text{ figure}}{\textbf{Area} \text{ of original figure}}$, we will write $\dfrac{A_{new}}{A_{original}}$.

a)   Copy the given figure on a sheet of dot paper, and then enlarge the original figure by making its corresponding sides three times as long.

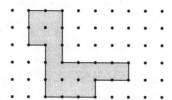

b)   Do it again making each corresponding side five times as long as in the original.

c)   Compute and reduce the ratios $\dfrac{S_{new}}{S_{original}}$, $\dfrac{P_{new}}{P_{original}}$, and $\dfrac{A_{new}}{A_{original}}$ for parts (a) and (b).

EF-20. Suppose you enlarge (or reduce) a figure on dot paper. Write what you think the ratios $\frac{P_{new}}{P_{original}}$ and $\frac{A_{new}}{A_{original}}$ will be if …

a) the ratio $\frac{S_{new}}{S_{original}}$ is $\frac{2}{1}$.

b) the new figure has corresponding side lengths three times the original.

c) the new figure has corresponding side lengths 10 times the original.

EF-21. Make a conjecture about what the ratios $\frac{P_{new}}{P_{original}}$ and $\frac{A_{new}}{A_{original}}$ will be if the ratio $\frac{S_{new}}{S_{original}}$ is $\frac{N}{1}$, where N represents a positive integer.

EF-22. Pat created a figure on dot paper with a perimeter of 54 units and an area of 90 square units. Kim reduced the figure so the side lengths are $\frac{1}{3}$ the original sides. Find the perimeter and area of the figure Kim made.

EF-23. ♠ ♦ ♥ ♣ Two aces have been removed from a deck of 52 playing cards.

a) Find the probability of drawing a King from the remaining deck.

b) Find the probability of drawing an ace from the remaining deck.

c) Write each of the probabilities as a percent.

EF-24. Here are some equations to solve for x. They look a bit different from the equations you've already solved, but you can use the same process to solve them that you've used before.

a) $\frac{x}{3} = 10$

b) $\frac{2x}{5} = 9$

c) $\frac{x}{3} = \frac{6}{7}$

d) $\frac{3x}{2} = \frac{12}{5}$

e) $\frac{5x}{8} = \frac{x-4}{4}$

EF-25.   René collected information from his fellow basketball players and made this graph:

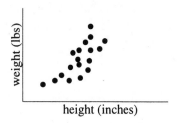

Describe what you notice about the data René plotted by copying and completing this sentence: In general, as height increases, _____.

EF-26.   Copy and solve these Diamond Problems:

    a)   b)

EF-27.   Rewrite each of the following products.

a)   6(x + 2)   c)   x(3 + 5)

b)   5(a + 3)   d)   y(6 + 4)

EF-28.   Marybeth wanted to multiply 32 by 14 without using a calculator, so she drew the following rectangle:

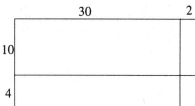

a)   Explain why the area of the large composite rectangle is equal to 32 · 14.

b)   Find a value for 32 · 14 by finding the areas of the four smaller rectangles and summing.

EF-29.   Write a ratio for each of the following descriptions.

a)   eight minutes to one day (in minutes)

b)   one gallon to one pint (in pints)

c)   forty-five minutes to two hours (in minutes)

d)   one pound to twelve ounces (in ounces)

EF-30. **Symbol**: " | | "
We will use the symbol " | | " to denote the length of a line segment.
For example, " |AB| " means "the length of segment AB."
In the diagram below we see that the segment AB has length 3, so we write |AB| = 3.
We also see that segment BD has length x + 2, so we write |BD| = x + 2.

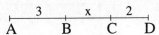

a) Copy the information about " | | " in your Tool Kit.

b) Write statements using the symbol " | | " for the length of segment AC and the length of segment AD.

EF-31. **Barbie Trivia** According to a July 2, 1996 article from the Davis Enterprise, the typical American girl between the ages of 3 and 11 owns an average of eight Barbie dolls. Barbie's popularity contributes to sales: in 1995, Mattel sold $1.4 billion worth of Barbie-related merchandise, up from sales of $430 million in 1987.

a) Find the difference in sales from 1987 to 1995, and write the result in both standard notation and scientific notation.

b) Find the ratio of 1995 sales to 1987 sales, and write the result in both standard notation and scientific notation.

4.3 SIMILAR TRIANGLES: MORE GEOMETRIC RATIOS

EF-32.   Ginger made the following enlargement of the "L-shaped" figure ABCDEF. Then Ginger gave her sketches to the Professor and asked if he could figure out the enlargement ratio.

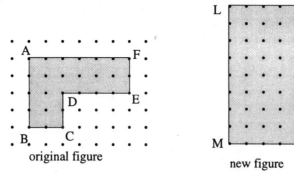

original figure

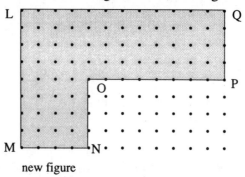

new figure

a)   The Professor used the ratio $\frac{|LQ|}{|AF|}$ to find the enlargement ratio $\frac{S_{new}}{S_{original}}$.
Name three more ratios of sides that he could have used, and then state the numerical value of the enlargement ratio.

b)   Always experimenting, the Professor decided to form some new ratios. Instead of choosing a ratio of the form $\frac{S_{new}}{S_{original}}$ of sides **between** the two figures, he picked ratios of sides **within** one figure. For example,

the ratio $\frac{|AB|}{|AF|}$ is of the form $\frac{\text{length of an original Side}}{\text{length of a second original Side}}$ and

$\frac{|LM|}{|LQ|}$ is of the form $\frac{\text{length of the corresponding new Side}}{\text{length of new Side corresponding to second original side}}$.

Find and reduce $\frac{|AB|}{|AF|}$, a ratio of sides from **within the original figure**.

Now find and reduce $\frac{|LM|}{|LQ|}$, a ratio of sides from **within the new figure**.

c)   Repeat part (b) for the ratio $\frac{|AB|}{|BC|}$ from **within the original figure** and the ratio $\frac{|LM|}{|MN|}$ from **within the new figure**.

d)   Repeat part (b) for the ratio $\frac{|AF|}{|DC|}$ from **within the original figure** and the ratio $\frac{|LQ|}{|ON|}$ from **within the new figure**.

EF-33.   In EF-32, how the ratios of sides **within** the original figure compare to the ratio of corresponding sides within the new figure? Write a note that the Professor could give to Ginger explaining any patterns you observed.

EF-34.   **Definition**: **similar** triangles
           We say two triangles are **similar** if their corresponding angles are the same size (that is, "congruent").
           For example, the two triangles shown below are similar since all the angles marked in the same way are the same size. Here, △ABC is similar to △DEF.

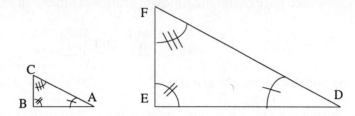

           Copy the definition for similar triangles (include the drawings) in your Tool Kit.

EF-35.   **Ratios of sides BETWEEN two similar triangles** Ginger noticed that when triangles are similar, it seems like one is an enlargement of the other. So, she decided to look for a relationship between the ratios of sides of similar triangles. Because they were handy, she used the triangles from EF-34, △ABC and △DEF.

           Since she was thinking of enlargements, Ginger decided to find the ratios of sides by choosing one side from each of the triangles, like she did when she was finding enlargement and reduction ratios.

   a)   Ginger measured the lengths of each of the sides of △ABC and △DEF. The measurements, rounded to the nearest millimeter, are given below. On a copy of the resource page for EF-35, label the sides of the triangles with the appropriate lengths.

           |AB| = 18 mm |BC| = 12 mm |AC| = 20 mm

           |DE| = 54 mm |EF| = 36 mm |DF| = 60 mm

           Write the following ratios of sides, and then reduce each ratio:

$$\frac{|AB|}{|DE|}, \ \frac{|AC|}{|DF|}, \ \text{and} \ \frac{|BC|}{|EF|}$$

           What do you notice about the three ratios?

   b)   Now look at △ABC and △JKL. How can you tell by looking at the sketch that the two triangles are similar?

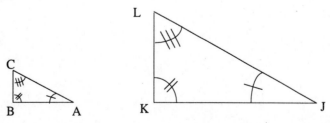

                      [PROBLEM CONTINUED ON NEXT PAGE]

EF-35. continued

c)   Ginger measured the sides of ΔJKL, too, so that she could compare more ratios. The measurements, rounded to the nearest millimeter, are as follows:

|JK| = 45 mm |KL| = 30 mm |JL| = 50 mm
On the resource page, write the following ratios of sides and reduce each ratio to lowest terms.

$$\frac{|AB|}{|JK|}, \; \frac{|AC|}{|JL|}, \text{ and } \frac{|BC|}{|KL|}$$

What do you notice about these ratios?

e)   On the resource page, use the measurements given in parts (a) and (c) to compare the ratios of corresponding sides of ΔDEF and ΔJKL by writing the following ratios of sides and reducing each to lowest terms. What do you observe?

$$\frac{|DE|}{|JK|}, \; \frac{|DF|}{|JL|}, \text{ and } \frac{|EF|}{|KL|}$$

EF-36.   a)   In EF-35, you compared only certain ratios of sides; that is, ratios of **corresponding** sides. How can you tell if sides of two similar triangles are corresponding?

b)   What did you observe about the ratios of the corresponding sides of two similar triangles? Write your observations in complete sentences.

EF-37.   Ginger told the Professor her conjecture that if two triangles are similar, then one is just an enlargement of the other. How can the Professor check Ginger's conjecture? Is ΔDEF an enlargement of ΔABC? If so, explain why, and find the enlargement ratio of the sides of ΔDEF to the corresponding sides of ΔABC. Otherwise, explain why not.

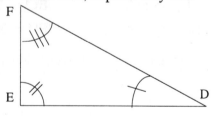

EF-38.   Triangles ABC and DEF are similar triangles.

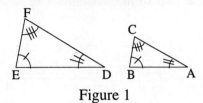

Figure 1

a)   Use Figure 1 to find two more ratios of sides which are equal to $\dfrac{|EF|}{|BC|}$.

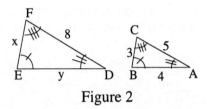

Figure 2

b)   Use the lengths of the sides given in Figure 2 and the ratios from part (a) to write an equation you can solve for x, the length of EF. Then solve your equation to find |EF|.

c)   Follow the process of parts (a) and (b) to write an equation to solve for y, the length of ED. Then solve your equation for y.

EF-39.   **Ratios of sides WITHIN similar triangles** Ginger needs your help to find a second ratio relationship for similar triangles. She decided to work with two similar triangles she had already measured, ΔABC and ΔDEF.

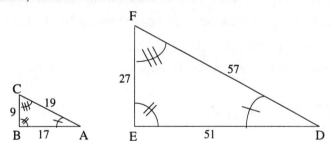

a)   First, from ΔABC, calculate the ratio $\dfrac{|AB|}{|BC|}$ to the nearest tenth. Then find a pair of sides from ΔDEF with a ratio equal to $\dfrac{|AB|}{|BC|}$.

b)   Now, calculate the ratios $\dfrac{|AB|}{|AC|}$ and $\dfrac{|BC|}{|AC|}$ to the nearest tenth. Which ratio of sides from ΔDEF will be equal to $\dfrac{|AB|}{|AC|}$? Which ratio of sides from ΔDEF will be equal to $\dfrac{|BC|}{|AC|}$? Make predictions, and then check them with your calculator.

c)   Write a couple of sentences to Ginger explaining the relationship of ratios of sides within similar triangles. You might want to include a sketch.

EF-40.   Suppose you are given these two similar triangles:

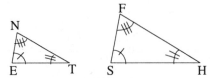

Look back at your results from EF-35 and EF-39. Use what you learned to make two conjectures about what is true about ratios of corresponding sides of $\triangle NET$ and $\triangle FSH$. Write your conjectures carefully and completely.

EF-41.   Maya and Mike are 60 miles apart and traveling towards each other. Maya travels at 20 miles per hour on her bike while Mike can manage 30 miles per hour in his car. In how many hours will they meet?

EF-42.   a)   Graph the equation $y = 2^x$.

b)   Using your graph, estimate to the nearest tenth for which x-values, the y-value is 5.

c)   Use a calculator to check if $5 = 2^x$ for the values of x you found in part (b).

d)   Use your calculator to see if you can find a value of x closer to the solution of $2^x = 5$.

EF-43.   Kara bought a box of beads in 13 different colors. There were four beads of each color (red, blue, green, orange, violet, …). Two red beads fell out of the box as Kara was going home.

a)   Find the probability of drawing a green bead from the beads that remain in the box.

b)   Find the probability of drawing a red bead from the beads that remain in the box.

c)   Write each of the probabilities as a percent.

d)   Compare your work in this problem with your work in EF-23. What do you notice?

EF-44.   Rewrite each of the following products.

a) $5(z + 4)$ b) $x(3 + 4)$

c) $x(x + 3)$ d) $y(y + 7)$

EF-45.   Sketch a geometric figure described by each of the following expressions, and then find each product using the area of each figure.

a) $(10 + 4)(10 + 3)$ b) $(10 + 1)(20 + 5)$

EF-46.   Rewrite each of the following expressions using each base only once.

a) $10^6 \cdot 10^9 \cdot 7^3 \cdot 7^4$ b) $x^3 \cdot x^3 y^2$

c) $3x^3 y \cdot 4x^2 y^2$ d) $(x^2)^3$

e) $(2x^2)^3$ f) $\dfrac{8x^2 y^3}{2x^2 y}$

EF-47.   **Slides: A Way to Represent Change**   Chauncey likes to play the slot machines in Lake
Tahoe. Last Tuesday, he started at 8 o'clock in the morning with $20. Lady Luck was with
him, and Chauncey played steadily for five hours. On counting his money, he realized that he
had made a gain of $40!

We can represent the change in Chauncey's pocket money with a graph as follows:

Chauncey starts the   *Five hours later, he is*   *He counts, and*   *Chauncey ends the*
day on Tuesday   *ready to count his*   *realizes he has gained*   *day after 5 hours with*
with $20:   *winnings:*   *$40:*   *$60 in his pocket:*
                              Slide 5 units to the
Start at point (0, 20).   right.   **Slide** 40 units up.   **End** at point (5, 60).

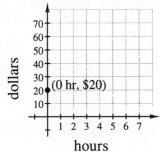 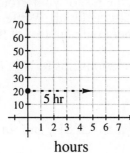 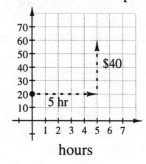 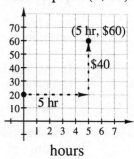

So, a **slide** is a way to record change graphically.

Describe how Chauncey's pocket money changed on Wednesday, represented by the following
graphs.

Wednesday

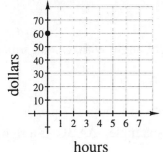 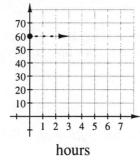 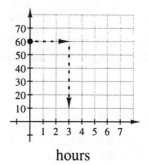 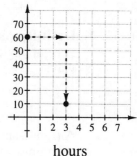

EF-48.   a)   Start with a point P at (−1, −5). Slide P three units right and then four units up. What
are the coordinates of P's new position?

b)   A **slide** is described only by the horizontal and vertical changes. How could you
describe the slide that was used to move point Q from (−3, 1) to (4, −2) ?

EF-49.   Solve each of the following equations for x.

a) $3x + c - 5 = -4$

b) $3c + x - 5 = -4$

c) $7(x - 2) = m + 3x$

4.4 MORE SIMILARITY: RIGHT TRIANGLES AND MISSING SIDES

EF-50.   Suppose you are given these two similar triangles:

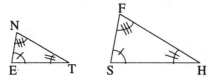

In EF-40, you wrote two conjectures about ratios of corresponding sides of ∆NET and ∆FSH. Compare your conjectures with the rest of your group members' and revise to come up with two group conjectures you are all satisfied with. Write your group conjectures carefully and completely, and be ready to share them with the rest of the class.

In this section, you will use these ratio relationships for similar triangles to find the length of an unknown side.

EF-51.   **Missing Sides** Triangle NET is similar to ∆FSH, and, in the drawing below, the lengths of some of the sides are given.

 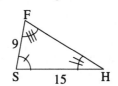

a)   Use the two conjectures from the previous problem to write two different equations you could use to solve for the length x of side NE.

b)   Find the length x using one of your equations.

EF-52.   **Definitions**: **right angle** and **right triangle**
A right angle is an angle that measures 90°. We indicate a right angle with a small square, this way:

    or   or

A right triangle is a triangle with exactly one right angle. For example, ∆ABC is a right triangle with its right angle at B.

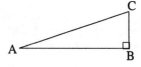

a)   Copy the definitions of "right angle" and "right triangle," including the diagrams, in your Tool Kit.

b)   There are four right triangles in the following sketch. Draw each one separately and write the name of each next to its drawing.

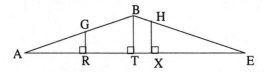

EF-53.   a)   Look at right triangles MAT and TCH below. How can you tell that they are similar?

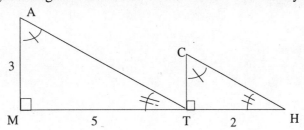

b)   Find |TC| by writing and solving an equation. (Hint: Use the variable x to represent the unknown length |TC|.)

c)   Write another equation using ratios that you could have used to solve for the length of side TC.

EF-54.   These two right triangles are similar:

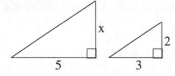

a)   Write <u>two</u> different equations you could use to solve for the length x.

b)   Use one of your equations to find the length x.

EF-55.   Write one equation based on the diagram below. Use your equation to solve for x.

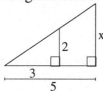

EF-56.   The two right triangles below are similar. Triangle DEF is an enlargement of △ABC.

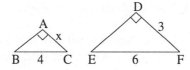

a)   Find x, the length of AC.

b)   What is the enlargement ratio?

c)   What do you think the ratio of the perimeter of △DEF to the perimeter of △ABC is?

d)   What do you think the ratio of the area of △DEF to the area of △ABC is?

EF-57.   Solve each of the following equations.

   a) $\frac{x}{3} = 7$

   b) $\frac{x}{3} = \frac{4}{7}$

   c) $\frac{x}{3} = \frac{x + 1}{7}$

   d) $\frac{3}{x} = \frac{4}{5}$

   e) $\frac{3}{x + 1} = \frac{8}{x}$

EF-58.   The length of a rectangular picture frame is twice the width. The perimeter is 132 centimeters. What is the width of the frame? (Hint: Draw a rectangle and label its sides.)

EF-59.   Rewrite each of the following products.

   a) $5(5 + 6)$ c) $x(5 + 6)$

   b) $5(x + 6)$ d) $x(x + 6)$

EF-60.   Dubious Dan is sure that $2(x + 4) = 2x + 4$. Doubtful Debbie is equally sure that $2(x + 4) = 2x + 8$.

   a)   Which claim is correct?

   b)   Use words and/or pictures to give an explanation that demonstrates the correct equation.

EF-61.   Plot the points $(0, 0)$, $(0, 12)$, $(34, 12)$, and $(34, 0)$ on regular graph paper. Use a ruler to connect the points to form a rectangle.

   a)   What are the length and the width of the rectangle?

   b)   To compute the area of the rectangle easily, divide it into four sections by thinking of the dimensions as $(10 + 2)$ by $(30 + 4)$. Then find the area of each section.

   c)   What is the area of the rectangle in square graph paper units?

EF-62.   Solve the following problem and write an equation that represents it.

   Find three consecutive odd numbers such that the sum of the smallest and seven times the largest is 68.

EF-63.   Find each sum without using a calculator. Leave your results in fractional form.

a) $\dfrac{3}{7} + \dfrac{4}{5}$

b) $\dfrac{3}{x} + \dfrac{4}{5}$

c) $\dfrac{x}{3} + \dfrac{x + 1}{7}$

EF-64.   Describe a **slide** used to move a point from $(3, 0)$ to $(5, 2)$. If you need help getting started, look at EF-47 and EF-48.

EF-65.   The moon is about 3.8×10^5 kilometers from the earth. If light travels at about 300,000 kilometers per second, how long does it take light from the moon to reach the earth?

EF-66.   The mass of the sun is about 330,000 times the earth's mass. If the mass of the earth is about $6 \cdot 10^{24}$ kilograms, what is the mass of the sun?

EF-67.   a)   Use a calculator to compute $(1.4 \cdot 10^{98}) \cdot (2.3 \cdot 10^5)$.

b)   Explain what happened.

c)   Now use what you know about exponents to compute $(1.4 \cdot 10^{98}) \cdot (2.3 \cdot 10^5)$.

EF-68.   Copy and solve these Diamond Problems:

EF-69.   Use the distributive property to rewrite each of the following expressions.

a) $2(x + 4)$ d) $3x + 6$ (Hint: $3x + 6 = 3 \cdot x + 3 \cdot 2$)

b) $x(x + 4)$ e) $4y - 8$

c) $x(x - 2)$ f) $m^2 + 5m$

EF-70.   Scientists use beakers of many shapes for holding and measuring liquids. Imagine that you
           add water at a steady rate to each of the beakers illustrated below. As water is added the height
           of the water in the beaker will rise. The graph shows how the height of the water increases as
           water is added at a steady rate to beaker A.

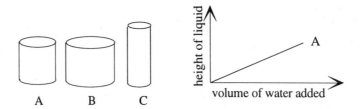

           a)   Copy the beaker shapes and the graph for A. Plot lines to show how beakers B and C
                  fill up as water is added. Label each line.

           b)   Explain why you drew the lines where you did.

4.5 USING EQUIVALENT RATIOS TO GRAPH

EF-71. **Fruit Punch** Lori's favorite punch recipe uses three cups of sugar for five gallons of punch. She wants to make seven gallons of punch for her daughter's gymnastics club party. She knows she could double the recipe to make ten gallons, but she doesn't want to make any extra punch. In this section, you will explore how geometric and algebraic ratios are related, and look at several ways to solve the problem of how much sugar to add.

    a) Make a graph of the line which represents this situation with "gallons of punch" on the horizontal axis. Explain why the points $(0, 0)$ and $(10, 6)$ lie on this line.

    b) Form a right triangle by drawing the vertical line $x = 10$, and label the lengths of the vertical and horizontal sides of the triangle. (Hint: One side of the triangle is on the x-axis.)

    c) Form a second right triangle within the first one by drawing the vertical line $x = 7$. Label the length of the horizontal side of the triangle. Since you don't know the length of the vertical side, label it y. What does y represent in the context of this problem?

    d) Use your graph to estimate how much sugar is needed for 7 gallons of punch.

    e) To get a more accurate answer, you can use an algebraic approach. First, explain why the two triangles from parts (b) and (c) are similar. Then, write an equation using equivalent ratios to find an exact value for y.

EF-72. a) Explain how you could use your graph from EF-71 to estimate the amount of sugar needed for 18 gallons of Lori's punch. What estimate do you get using your graph?

    b) Write an equation and solve the problem in part (a) algebraically.

EF-73. Marta has 8.5 cups of sugar in the pantry. If she uses it all in the same sugar-to-punch ratio as in Lori's recipe, how much punch can she make?

    a) Use your graph to estimate the number of gallons of punch.

    b) Write an equation and solve the problem in part (a) algebraically.

EF-74.   C.J.'s car gets 20 miles per gallon of gas.

a)   Copy and complete the following table of data for C.J.'s car:

# miles traveled	# gallons of gas used
20	
100	
40	
10	
0	
m	

b)   Use the table you made in part (a) to write an equation relating g, the number of gallons used, to m, the number of miles driven. A good way to check your equation is to answer the question: Does your equation make sense for m = 20 miles?

c)   Graph your equation with m, the number of miles, on the horizontal axis, and g, the number of gallons, on the vertical axis. Label both axes. Scale the horizontal axis so one unit represents 10 miles. Scale the vertical axis so one unit represents one gallon.

d)   Use your graph to estimate the value for g when m = 72.

e)   Now form two similar triangles by drawing the vertical lines m = 20 and m = 72. Explain why these triangles are similar, and label the lengths of the horizontal and vertical sides of each triangle. (You will have to use a variable for one of the lengths.)

f)   Use your triangles to write an equation with two equivalent ratios, and solve it to find out how much gas C.J. used in driving 72 miles. Compare your solution to your estimate in part (d).

For problems EF-75 and EF-76, write an equation and solve it.

EF-75.   Kim noticed that 100 vitamins cost $1.89. At this rate, how much should 350 vitamins cost?

EF-76.   Joe came to bat 464 times in 131 games. At this rate, how many times may he expect to come to bat in a full season of 162 games?

EF-77.   In a city of three million, 3472 people were surveyed. Of those surveyed, 28 of them admitted that they watched "Gilligan's Island."

a)   What fraction of the people surveyed admitted watching the program?

b)   What percentage of the people surveyed admitted watching the program?

c)   If the survey did represent the city's TV viewing habits, how many people in the city would admit watching the program?

EF-78.   A $5'' \times 7''$ photograph is enlarged so its width is 17.5 inches.

   a)   What is the length of the enlarged photograph? (Note that the width of the original photo is $5''$.)

   b)   What is the enlargement ratio?

EF-79.   L.J.'s car has a gas tank which holds 19 gallons.

   a)   If L.J. used eight gallons to drive 200 miles, does the car have enough gas to go another 250 miles?

   b)   What assumptions did you make in your solution to part (a) ?

EF-80.   Solve each of the following equations.

   a) $\frac{x}{3} = \frac{5}{7}$ b) $\frac{5}{y} = \frac{2}{y+3}$

   c) $\frac{1}{x} = \frac{5}{x+1}$ d)   Draw two similar triangles for which the equation in part (a) can be used to find a side length x.

EF-81.   Is it true that $3^4 \cdot 3^5 = 3^9$? Justify your answer by using numbers, or diagrams, or some other method you choose.

EF-82.   Write and solve an equation using ratios to find $|BY|$.

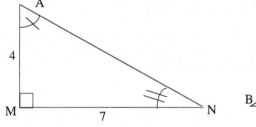
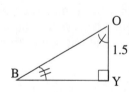

4.6 WRITING AND SOLVING EQUATIONS INVOLVING RATIOS

For each of problems EF-83 through EF-85, write an equation and solve it.

EF-83.   Two numbers are in a ratio of 1:5. If their sum is 30, find the numbers.

EF-84.   Two numbers are in a ratio of 2:3. If their sum is 50, find the numbers.

EF-85.   Tina has 24 ounces of gas and one ounce of oil in her motorcycle. How much oil must be added to make the ratio of gas to oil 16:1 ?

EF-86.   Larry weighs 240 pounds and decides to go on a miracle diet. The diet claims he will lose 20 pounds in three days.

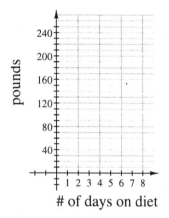

a)   Use a **slide** to show the change Larry could expect if the diet's claim is true. Describe the slide in a sentence.

b)   What would Larry's weight be after three days if the claim is correct?

c)   Predict Larry's weight in six days.

d)   Predict Larry's weight in 12 days.

e)   Suppose Larry stuck to the diet and kept losing weight at the constant rate of 20 pounds every three days. When would Larry disappear completely?

f)   As you figured in part (e), it doesn't make sense for Larry to keep losing weight at a constant rate change. Express the rate at which Larry is losing weight as a ratio. Realistically, what would happen to this ratio as time passed?

EF-87.   A line contains the points (3, 1) and (4, 3). Find the coordinates of five other points on the line.

EF-88.   A number y is 14% of 300.

a)   Use equivalent ratios to write an equation that expresses this fact.

b)   Solve your equation to find y.

EF-89.   The number 54 is 16% of z.

a)   Use ratios to write an equation that expresses this fact.

b)   Solve your equation to find z.

EF-90.   Use ratios to write an equation that can be used to answer each of the following questions and then solve the equation.

a)   What number is 25% of 40 ?

b)   Twenty-five is what percent of 40 ?

c)   Twenty-five is 40% of what number?

EF-91.   A biscuit recipe uses $\frac{1}{2}$ teaspoon of baking powder for $\frac{3}{4}$ cup flour. How much baking powder is needed for three cups of flour?

EF-92.   The length of a rectangle is six centimeters more than the width. If the ratio of the length to the width is five to three, find the rectangle's dimensions.

EF-93.   Malwart's sells posters for $15.00. The store's profit on each poster is $6.00. Write a ratio of the profit to the selling price. What percent of the selling price is the profit?

EF-94.   The local art supply store charges 98¢ tax on a $12.00 purchase. Write a ratio of the tax to the purchase price. What percent of the purchase price is the tax?

EF-95.   Describe a slide that could be used to move a point from (0, –2) to (–3.5, 5).

EF-96.   Suppose that ⊙ is a mystery operation that works as follows:

$$4 \odot 3 = 10$$
$$4 \odot 7 = 18$$
$$7 \odot 4 = 15$$
$$0 \odot 7 = 14$$

a)   Look for a pattern above and use it to solve $3 \odot 5 = y$ and $6 \odot x = 10$.

b)   Write a rule to describe how the operation ⊙ works in general: $a \odot b = \underline{?}$.

EF-97.   Suppose you have a penny, a nickel, and a dime and you toss them onto the ground.

a)   Find the probability that the penny comes up 'heads.'

b)   Find the probability that both the penny and the dime come up heads.
        (Hint: It helps to make a table of the eight possible outcomes.)

c)   Find the probability that exactly two of the coins come up heads.

   d)   Write the probabilities you found in parts (a), (b), and (c) as percents.

EF-98.   Use the distributive property to rewrite each of the following expressions.

   a) $4(y - 7)$ d) $5m + 10 = 5 \cdot m + 5 \cdot 2 =$

   b) $2y(y + 4)$ e) $7y + 49$

   c) $3z(2z - 4)$ f) $m^2 + m$

EF-99.   Copy and solve these Diamond Problems:

EF-100.   Use a calculator to find each of the following products. Write each result in scientific notation.
          For example, $0.02 \cdot 0.1 = 2.0 \cdot 10^{-3}$.

   a) $0.00125 \cdot 0.02$

   b) $3 \div (1.5 \cdot 10^{-4})$

   c) $0.0738 \cdot (6.2 \cdot 10^{17})$

   d) $360{,}000{,}000{,}000 \div (3 \cdot 10^{14})$

EF-101.   Write an expression for the perimeter and the area of each of the rectangles below.

   a)   b)

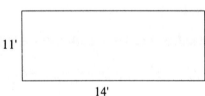

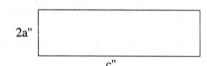

EF-102.   a)   Use subproblems to find the perimeter and area of this figure:

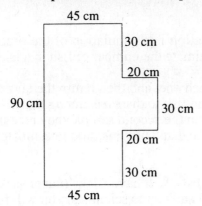

b)   Find the lengths of the unlabeled sides in the figure below. Then find the perimeter and area of the figure.

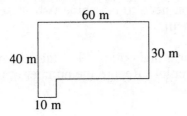

4.7 ESTIMATING FISH POPULATIONS: A SIMULATION

EF-103.   **Estimating Fish Populations** This investigation is a simulation of the process used by Fish and Game scientists to mathematically estimate the number of fish in a lake.

Scientists net some fish, count them and tag each one, and then return the tagged sample to the lake. After allowing enough time for the tagged fish to disperse among the other fish in the lake, the scientists net another sample of fish. In the second sample they note the ratio of the number of tagged fish to the number of fish netted and use this ratio to estimate the total fish population of the lake.

Your group, acting as a team of research scientists, will have a lake (paper sack) full of fish (beans), and a net (a small cup). Just like Fish and Game scientists, you will be trying to estimate the number of fish in your lake by netting samples and using ratios.

a)   Before you start your mathematical investigation, *guess* how many fish are in your lake. Write down your guess to compare later with the actual number.

b)   **Tagging the Fish** In your first outing, you need to tag some fish. Use your "net" to take out a cupful of fish in order to "tag" them.

c)   Count the number of fish you netted in your initial outing. To "tag" the netted fish, replace each one with a bean of a different color. Record the number of tagged fish as the **"total number of tagged fish."**

d)   Put all the tagged fish back into the lake. (Put the beans they replaced in a "holding pond.") Be careful not to let any of the fish jump out onto the floor. Gently shake the bag to thoroughly mix all the fish in the lake. Try not to bruise them.

e)   **First Sample** Use your net to take out another cupful of fish. Count the number of tagged fish in the sample and record this number as **"number of tagged fish in sample."** Count and record the **"total number of fish in sample,"** also.

f)   You now have three pieces of information: the **total number of tagged fish** in the lake, the **number of tagged fish in the sample**, and the **total number of fish in the sample**. Use this information to write two equivalent ratios. Then use the equation you wrote to find a mathematical estimate of the total number of fish in the lake.

g)   **Second Sample** Return your sample of fish to the lake, gently mix the fish, and take another sample. Repeat the counting procedure of part (e) and the use of ratios in part (f) to get another estimate of the lake's fish population.

h)   It is important to get an accurate count of the fish population, but each time you take the boat out to net fish it costs the taxpayers for your time and equipment. It cost $1000 to net and tag the fish initially, and each time you net an count a sample it costs $800. So far your three outings have cost $2600. If you feel your estimate at this point is accurate, record it on the class chart with your cost. If you think you should try another sample for better accuracy, do the same steps as before. Draw as many samples as you feel you need, but remember each sampling costs $800.

[PROBLEM CONTINUED ON NEXT PAGE]

EF-103. continued

    i)   After your data is recorded on the class chart, count the fish in your lake to find the actual population. (Biologists can't actually do this!) Record the actual population on the class chart.

         How close was your estimate? Calculate the percentage error. What might have thrown your estimate off? What do you think of this method for estimating fish populations?

    Be sure each group member records all the group's data, calculations, and conclusions.

EF-104.   **At The Movies** It is now 7:51 PM. The movie that you've waited three weeks to see starts at 8:15. Standing in line for the movie, you count 146 people ahead of you. Nine people can buy their tickets in 70 seconds. To determine if you will be buying your ticket before the movie starts, write a ratio equation and solve it.

EF-105.   A 70-foot length of wire weighs 23 pounds.

    a)   How much does zero feet of wire weigh?

    b)   Draw a graph showing the weight of the wire (vertical axis) compared to the length of the wire (horizontal axis). Label the axes with appropriate units.

    c)   Use your graph to **estimate** the weight of a 50-foot length of wire. Then, to find a more accurate answer, write and solve an equation using equivalent ratios to find the weight of a 50-foot length of wire.

    d)   Use your graph to estimate the length of 15 pounds of wire. Then, to find a more accurate answer, write and solve an equation using equivalent ratios to find the length of a piece of wire that weighs 15 pounds.

    e)   How close were your estimates to the actual values you found?

EF-106.   Solve the following problem and write an equation that represents it.

    Admission to the football homecoming dance was $3 in advance and $4 at the door. There were 30 fewer tickets sold in advance than at the door and the ticket sales totaled $1590. How many of each kind of ticket were sold?

EF-107.   Use what you know about exponents to rewrite each of the following expressions.

    a) $3x^2 \cdot x$ b) $\dfrac{n^{12}}{n^3}$ c) $(x^3)^2$

    d) $(-2x^2)(-2x)$ e) $\dfrac{-8x^6y^2}{-4xy}$ f) $(2x^3)^3$

EF-108.   Here's another beaker problem. The graph below shows how the height of the water increases as water is added at a steady rate to beaker A.

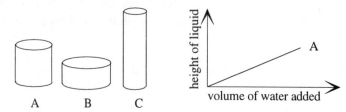

a)   Copy the beaker shapes and the graph for A. Plot lines to show how beakers B and C fill up as water is added. Label each line.

b)   Explain why you drew the lines where you did.

EF-109.   **Chapter 4 Summary: Rough Draft** These are the main ideas of Chapter 4 :

1.   The enlargement and reduction of shapes involves geometric ratios. You worked with the ratios of corresponding sides, the ratios of perimeters, and the ratios of areas of different figures.

2.   The ratios of corresponding sides of similar right triangles are equal.

3.   Ratio problems can be represented by graphs of lines passing through the origin.

4.   Equivalent ratios can be used to write equations to solve problems.

5.   Percent problems are ratio problems.

Write your answers to the following questions in rough draft form **on separate sheets of paper**, and be ready to discuss them with your group at the next class meeting. Focus on the **content**, not neatness or appearance, as you write your summary. You will have the chance to revise your work after discussing the rough draft with your group.

a)   From your homework assignments, select and recopy at least one problem that best illustrates your understanding of each of these main ideas. Tell why you selected the problems that you did. If you select one problem that demonstrates several of the main ideas, explain how it does this.

b)   If you had to choose a favorite problem from the chapter, what would it be? Why?

c)   Find a problem that you still cannot solve or that you are worried that you might not be able to solve on a test. Write out the question and as much of the solution as you can until you get to the hard part. Then explain what you think is keeping you from solving the problem. Be clear and precise.

4.8 SUMMARY AND REVIEW

EF-110.   **Chapter 4 Summary: Group Discussion** Take out the rough draft summary you completed in EF-109. Take this time to discuss your work; use homework time to revise your summaries as needed.

For each of the five main ideas of the chapter, choose one member of the group to lead a short discussion. The discussion leaders should take turns to:

- explain the problem they chose to illustrate their main idea,

- explain why they chose that particular problem, and

- explain as far as possible the problem they still cannot solve (part c).

This is your chance to make sure your summary is complete, update your Tool Kits, and work together on problems you may not be able to solve yet.

EF-111.   Suppose a right triangle with sides of lengths 5, 12, and 13 centimeters is similar to a right triangle whose shortest side is 15 centimeters long.

a)   What is the perimeter of the larger triangle?

b)   What is the ratio of the perimeter of the smaller triangle to the perimeter of the larger triangle?

c)   How does the ratio in part (b) above compare to the ratio of the lengths of corresponding sides of the triangles?

d)   What is the ratio of the area of the smaller triangle to the area of the larger triangle?

EF-112.   Carrie Ann's car gets 20 miles per gallon. For each question below, use equivalent ratios to write an equation, then solve it.

a)   How far will Carrie Ann's car go on 8 gallons of gas?

b)   If Carrie Ann drives 80 miles, how much gas will be used?

c)   If she drives 118 miles, how much gas will be used?

EF-113.   Mario's car needs 12 gallons of gas to go 320 miles.

a)   How much gas is needed to go 0 miles?

b)   Draw a graph that shows g, the number of gallons of gas used (vertical axis), compared to m, the number of miles driven (horizontal axis).

c)   From your graph, **estimate** how much gas will be needed to go 70 miles.

d)   **Estimate** how many miles can be driven with 10 gallons of gas.

e)   Use equivalent ratios to write equations to find the exact answers in parts (c) and (d). Solve your equations.

EF-114.   A certain line contains the points (3, 1) and (6, 2). Find the coordinates of five other points on the line.

EF-115.   a)   Enlarge this figure on dot paper by making each side of the new figure twice as long as its corresponding side in the original figure.

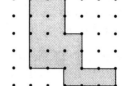

b)   Write the enlargement ratio of the corresponding sides.

c)   For the figures in part (a), write and reduce the ratios

$$\frac{\text{Length of \textbf{Side} of \textbf{new} figure}}{\text{Length of \textbf{Side} of \textbf{original} figure}},$$

$$\frac{\text{\textbf{Perimeter} of \textbf{new} figure}}{\text{\textbf{Perimeter} of \textbf{original} figure}}, \quad \text{and} \quad \frac{\text{\textbf{Area} of \textbf{new} figure}}{\text{\textbf{Area} of \textbf{original} figure}}.$$

EF-116.   Find |DE| and |DF|.

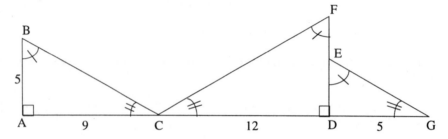

EF-117.   Find each of the following ratios.

a)   the value of one dime to the value of one quarter

b)   twelve feet to nine yards

c)   sixty ounces of M&M's to two pounds

EF-118.   Two numbers are in a ratio of 3:7 to each other. Their sum is 700. Find the two numbers.

EF-119.   Solve each of the following equations for x.

a) $\dfrac{x}{3} = \dfrac{x + 4}{5}$

b) $\dfrac{1}{x} = \dfrac{3}{x + 5}$

c) $\dfrac{17}{x} = \dfrac{25}{100}$

EF-120. Use the distributive property to rewrite each of the following products.

a) $5(x + 2)$

b) $4(3y - 2)$

c) $2t(3t + 5)$

d) $5(2t^2 - 3t + 1)$

e) $6(4x^{10} + 6x^3 + x)$

f) $(10 + 3)(20 + 7)$

EF-121. For each of the questions below, write an equation and solve it.

a) What percent of 60 is 45 ?

b) What is 45% of 60 ?

c) Sixty percent of what number is 45 ?

EF-122. Write each of the expressions below in a simpler exponential form.

a) $x^2y^3 \cdot x^3y^4$

b) $-3x^2 \cdot 4x^3$

c) $(x^3)^4$

d) $(2x^2)^3$

e) $\dfrac{6x^2y^3}{2xy}$

f) $(x^3y)^2(2x)^3$

EF-123. **Chapter 4 Summary: Revision** This is the final summary problem for Chapter 4. Using your rough draft from EF-109 and the ideas you discussed in your groups from EF-110, spend time revising and refining your Chapter 4 Summary. Your presentation should be thorough and organized, and should be done on a separate piece of paper.

Dot Paper

EF-4. Reducing Figures

a)

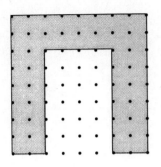

b)

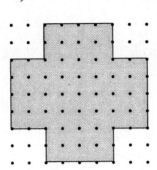

c)

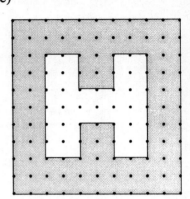

Dot Paper

Dot Paper

Algebra Tool Kit: the "what to do when you don't remember what to do" kit

EF-35. Ratios of sides between two similar triangles

Ginger is looking for a relationship between the ratios of corresponding sides of similar triangles ΔABC and ΔDEF.

- First, label each side with its length, as measured by Ginger. Then write the ratios of the side lengths, and finally reduce each ratio. What do you notice about the reduced ratios?

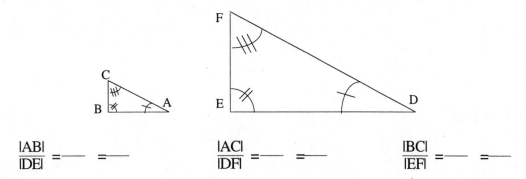

$$\frac{|AB|}{|DE|} = \underline{\quad} = \underline{\quad} \qquad \frac{|AC|}{|DF|} = \underline{\quad} = \underline{\quad} \qquad \frac{|BC|}{|EF|} = \underline{\quad} = \underline{\quad}$$

- Try it with another pair of triangles. Label the lengths of the sides the triangles, write the ratios of the side lengths, and then reduce each ratio. What do you notice about the reduced ratios?

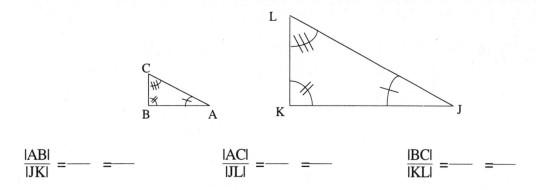

$$\frac{|AB|}{|JK|} = \underline{\quad} = \underline{\quad} \qquad \frac{|AC|}{|JL|} = \underline{\quad} = \underline{\quad} \qquad \frac{|BC|}{|KL|} = \underline{\quad} = \underline{\quad}$$

- Try it with a third pair of triangles. Label the lengths of the sides the triangles, write the ratios of the side lengths, and then reduce each ratio. What do you notice about the reduced ratios?

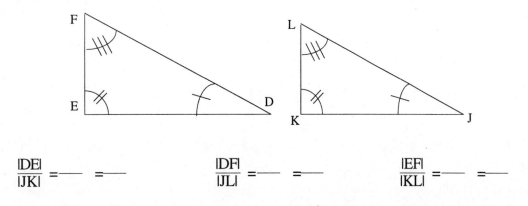

$$\frac{|DE|}{|JK|} = \underline{\quad} = \underline{\quad} \qquad \frac{|DF|}{|JL|} = \underline{\quad} = \underline{\quad} \qquad \frac{|EF|}{|KL|} = \underline{\quad} = \underline{\quad}$$

Chapter 5

TILING
RECTANGLES:
Factoring
Quadratics

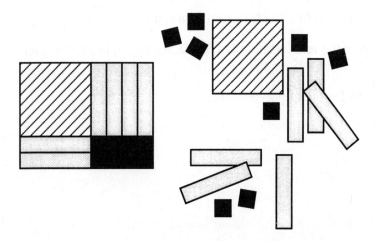

CHAPTER 5

TILING RECTANGLES:
FACTORING QUADRATICS

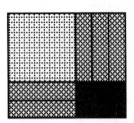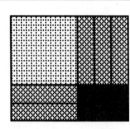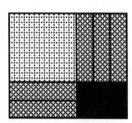

This chapter is based upon the fact that you can calculate the area of a rectangle by multiplying its length times its width. You sometimes will use normal measurement units, sometimes graph paper units, and other times variables to represent the lengths, widths, and areas of rectangles.

Algebra tiles, like those pictured on the cover of this chapter, will provide a handy device for figuring out many of the problems.

In this chapter, you will have the opportunity to:

- view multiplication and factoring of polynomials geometrically based on areas of rectangles;

- write expressions for the total area of a rectangle in two different but equivalent ways, as the sum of the areas of the individual parts, or as the product of the length and the width: "area as a product = area as a sum";

- see the inverse relationship between factoring and multiplying --

    use multiplication to change an expression from a product to a sum, and

    use factoring to change an expression from a sum to a product;

- explore the relationship between factoring a quadratic expression and solving a related quadratic equation; and

- add and subtract polynomials.

CHAPTER CONTENTS

5.1 TILING RECTANGLES

TR-1.   **Building Composite Rectangles** You can use algebra tiles to build **composite** rectangles. A composite rectangle is a rectangle made with at least two tiles. For example, you can build a composite rectangle using one large square, two rectangles, and one small square like this:

In each of parts (a) through (f), try to build a composite rectangle using **one** large square, **at most ten** rectangles, **and the number of small squares indicated**. Sketch each composite rectangle you build.

a)   one small square

b)   two small squares

c)   three small squares

d)   four small squares

e)   five small squares

f)   six small squares

g)   Now make a chart with the headings shown below. Include all the composite rectangles created by your group and all additional composite rectangles reported by other groups.

Number of Large Squares	Number of Rectangles	Number of Small Squares	Sketches
1	2	1	

TR-2.   Maxine wants to find a rectangle with an area of 24 square centimeters. Help her by using centimeter graph paper to draw as many different rectangles as you can (at least five) with area 24 square centimeters.

TR-3.   Plot the points (5, 4), (–3, 4), (–3, –2), and (5, –2) on regular graph paper. Use a ruler to connect the points to form a rectangle.

    a)   What are the length and the width of the rectangle?

    b)   What is the area of the rectangle in square graph paper units?

TR-4.   Plot the points (5, 3), (–4, 3), (–4, –6), and (5, –6) on a set of axes. Use a ruler to draw the quadrilateral that has these points as vertices.

    a)   What kind of quadrilateral was formed?

    b)   How long is each side of the quadrilateral?

    c)   What is the area of the quadrilateral in square graph paper units?

TR-5.   Dean found the area of the following rectangle using subproblems: he added the area of each part to find the total area, as shown below.

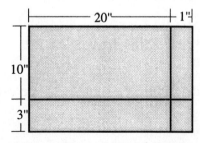

Dean's expression: Area = 100 + 60 + 20 + 12 = 192

Use multiplication to write a different expression for the area of the rectangle, and then multiply to check Dean's work.

TR-6.   Derek found the area of the following rectangle by figuring out the length and the width and then multiplying to find the total area.

Derek's expression: Area = (10 + 3)(20 + 1) = 13·21 = 273

Write a different expression for the area (as the sum of the parts) and then use it to check Derek's work.

TR-7.   Write each of the following expressions in a simpler exponential form.

a) $x^3 \cdot x^4$

b) $x^{-3} \cdot x^{17}$

c) $\dfrac{n^{17}}{n^{12}}$

d) $\dfrac{n^5}{n^{-2}}$

e)   If $x \neq 0$, we know that $\dfrac{x^3}{x^3} = 1$. But if we subtract exponents, we get $\dfrac{x^3}{x^3} = x^0$. What must be true about x^0 and 1 ?

TR-8.   a)   Show how you know that $\triangle ADE$ is an enlargement of $\triangle ABC$.
(Hint: What is the enlargement ratio for corresponding sides?)

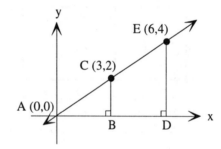

b)   What is the ratio of the areas of $\triangle ADE$ and $\triangle ABC$?

c)   Describe the slide from point A to point C.

d)   Describe the slide from point A to point E.

TR-9.   Write an equation and solve the following problem.

Renting a car costs $25 per day (even if you only use the car for part of the day) plus $0.06 per mile. What is the greatest number of miles that you can drive if you only have $40 ?

5.2 ALGEBRA TILES AND COMPOSITE RECTANGLES

TR-10.   **Algebra Tiles** The set of algebra tiles you used in TR-1 includes tiles of three types: large squares, rectangles, and small squares.

     a)   Suppose the large square has a side length of x units and the small square has a side length of 1 unit. Use this information to find the area of each of the tiles.

            the small square: the rectangle: the large square:

     b)   In your Tool Kit, trace each type algebra tile and mark the dimensions along each side. Clearly label each tile with its area. From now on, we will refer to each type of algebra tile by its area.

TR-11.   **Composite Rectangles and Area** Determine if it is possible to build a composite rectangle with each set of tiles indicated in the chart below. If so, on your Composite Rectangles and Area resource page sketch a composite rectangle you could build; if not, answer "No." For each composite rectangle, label each tile with its area as shown. Use this to write the area of the entire composite rectangle as a sum of the areas of the pieces.

At this point, we should start using the standard arrangement for our composite rectangles. Conventionally, the large squares are arranged in the upper left corner and the small squares are located in the lower right corner. The rectangles fill in the figure to form a composite rectangle.

Number of x^2 tiles	Number of 1x tiles	Number of 1 tiles	Is a composite rectangle possible?	Sketch	Algebraic Expression for the Area of the Composite Rectangle
1	3	2	Yes	x^2 1x 1x 1x 1 1	$x^2 + 3x + 2$
1	5	3			
1	4	4			
1	6	5			
1	3	9			
1	4	3			
1	7	10			

TR-12.   A cyclist travels for three hours at n miles per hour. He then decreases his speed by two
          miles per hour and cycles for two more hours. At the end, he has cycled 86 miles. Write an
          equation and solve for n.

TR-13.   For each of the following equations, make a table with at least five entries and draw a graph.

          a)   y = 0.75x + 2

          b)   y = –0.75x + 2

          c)   Describe two things that are the same and one thing that is different about the two graphs.

TR-14.*   Which of the activities listed below will
          produce a speed-time graph like the one
          shown?

          Select the activity you think fits best and
          explain exactly how it fits the graph.
          Write reasons why you reject the other
          activities.

          Biking   Golf
          Javelin Throwing   High Jumping
          Pole Vaulting   Water Skiing
          100 meter Sprint   Drag Racing
          Sky Diving   Skateboarding

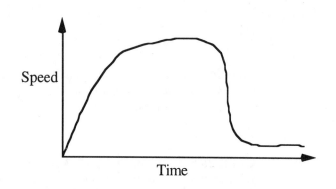

TR-15.   Fifty sophomores and 100 juniors want to take a computer graphics class, but the lab can
          accommodate only 50 people. The decision was made to take 20 sophomores and 30
          juniors. Names are drawn at random for each level.

          a)   If you were a sophomore who wanted to take the class, what is the probability you would
                be selected?

          b)   What is the probability of being selected if you were a junior?

          c)   Who has a better chance of being selected, someone who is a sophomore or someone
                who is a junior? Explain.

TR-16.   Solve each of the following equations.

          a) $14 = 8 + 5q$

          b) $\dfrac{12}{x} = 4$

          c) $s + 2(s + 1) + 3(s + 2) = 10s - 8$

          d) $w(w - 3) = 15 + w^2$

---

* Adapted from *The Language of Functions and Graphs*, Joint Matriculation Board and the Shell Centre for Mathematical
  Education, University of Nottingham, England.

TR-17.   Use the distributive property to write an equivalent expression for the area of each figure.

a)   Area = 4(x + 1)   Area = ?   b)   Area = 3(x) + 3(4)   Area = ?

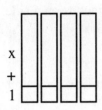 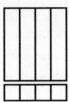   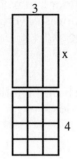 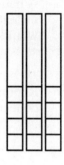

TR-18.   Solve each equation for y.

a) $5 + 2y = y - 8$

b) $x + 2y = y - 8$

TR-19.   Use exponents to write each of the following expressions as simply as possible.

a) $(5x^2)^2$

b) $\left(\dfrac{3x^2y}{x}\right)^3$

c) $\dfrac{x^5}{x^5}$

d)   For $x \neq 0$, explain how you know that $x^0 = 1$. If you need help, look back at problem TR-7.

TR-20.   While playing a dice game with her daughter, Julie figured out that the probability of rolling a total of seven points with two standard dice is $\frac{1}{6}$.

a)   What is the percentage that corresponds to $\frac{1}{6}$?

b)   If you rolled two standard dice 523 times, about how many times could you expect to roll a total of seven points?

5.3 REPRESENTING TILES ALGEBRAICALLY

TR-21.   A bathroom floor can be covered with 40 square tiles which are nine inches on a side. If instead the floor is covered with square tiles which are three inches on a side, how many tiles would be needed?

TR-22.   Represent each of the following tile collections with an algebraic expression. For example, this collection of tiles

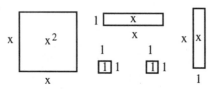

can be represented by the expression $x^2 + 2x + 2$.

a)   b)   c)   38 small squares
                                                                                    20 rectangles
                                                                                    5 large squares

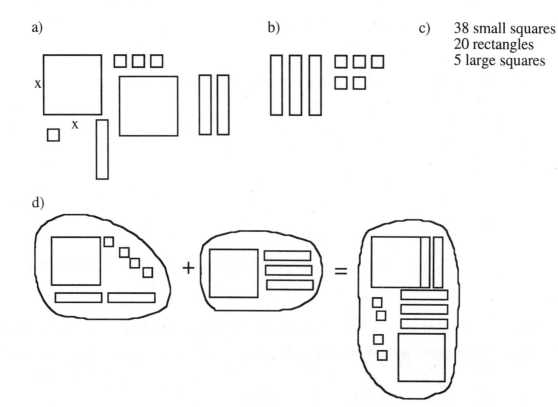

d)

TR-23.   Suppose one person in your group has two big squares, three rectangles, and one small square on his desk and another person has one big square, five rectangles, and eight small squares. You decide to put all the tiles together on one desk. Write an algebraic equation that represents this situation. Your equation should look like:

   (*tiles on his desk*) + (*tiles on another desk*) = *all tiles together on one desk.*

TR-24.   Bob, Chris, Janelle, and Pat are in a group. Bob, Chris, and Janelle have algebra tiles on their desks. Bob has two big squares, four rectangles, and seven small squares; Chris has one big square and five small squares; and Janelle has 10 rectangles and three small squares. Pat's desk is empty. The group decides to put all of the tiles from the three desks onto Pat's desk. Write an algebraic equation that represents this situation.

TR-25.   Imagine you were busy working on a problem with algebra tiles. You had three big squares, five rectangles, and 10 small squares on your desk when your friend leaned over and borrowed two big squares, two rectangles, and four small squares. Write an algebraic equation that represents the tiles you had, what your friend took, and the tiles you had left.

TR-26.   Suppose in class you were working with one big square, seven rectangles, and six small squares when you accidentally knocked the big square and five of the small squares onto the floor. Write an equation to represent the tiles you had, what fell off the desk, and the tiles that remained.

TR-27.   You were minding your own business using your algebra tiles – two big squares, four rectangles, and three small squares — when Tom came in and took the four rectangles and two of the small squares. Write an equation to represent the tiles you started with, what Tom took, and the tiles you had after Tom left.

TR-28.   One way to demonstrate that $2x$ does not usually equal x^2, is to use two rectangles and one large square as shown.
We say "$2x$ *does not usually* equal x^2" because *sometimes it does*. Check specifically that if $x = 0$, or if $x = 2$, then $2x = x^2$.

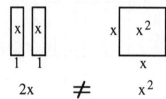

a)   Sketch the tiles you would use to show that $3x + x$ does not usually equal $3x^2$.

b)   Make a sketch of the tiles you would use to show that $2x - x$ does not usually equal 2.

TR-29.   Write an algebraic expression for the area of each of the following composite rectangles in two different ways — first as a product, and then as a sum. Use "x^2" to represent the large square, "x" to represent each rectangular tile, and "1" to represent each small square. For example, for this composite rectangle

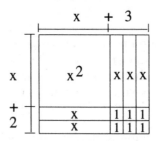

we can write

$$(x + 3)(x + 2) \quad = \quad x^2 + 5x + 6$$

area as a product   area as a sum

a)   b)   c)

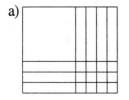   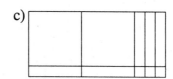

d)   e)

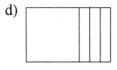   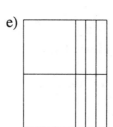

TR-30.   You know that the area of a rectangle can be found by multiplying its length and width:

$$\text{area} = (\text{length})(\text{width}).$$

You can use this fact to show that $x^2 + 5x + 6 = (x + 3)(x + 2)$ by arranging the appropriate tiles into a composite rectangle:

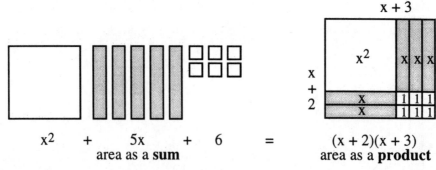

$$x^2 \quad + \quad 5x \quad + \quad 6 \quad = \quad (x + 2)(x + 3)$$

area as a **sum**   area as a **product**

Use algebra tiles to show why each of the following equations is true. Make a drawing to represent each equation and label each part.

a) $x^2 + 7x + 6 = (x + 6)(x + 1)$ b) $x^2 + 4x + 4 = (x + 2)(x + 2)$

c) $x^2 + 3x + 2 = (x + 2)(x + 1)$

TR-31. **Polynomials** In the previous problem, you saw that $x^2 + 5x + 6$ is equal to $(x + 2)(x + 3)$, both expressions represent the same area. The expression $x^2 + 5x + 6$ is an example of a "polynomial" (from the Greek words *poly* meaning *many*, and *nomos* meaning *part*).

A polynomial is a **sum** of "monomials" or "terms"; in this case, $x^2 + 5x + 6$ is the sum of the monomials x^2, $5x$, and 6. The expressions $3x^2$, $-15x$, x, and -7 are also examples of monomials (from the Greek words *monos* meaning *single*, and *nomos* meaning *part*).

Certain polynomials can also be written as products of factors. For example, the sum $x^2 + 5x + 6$ can be written as the **product** of the two **factors** $x + 2$ and $x + 3$. We say we have **factored** the polynomial $x^2 + 5x + 6$ as "the quantity $x + 2$ times the quantity $x + 3$."

Identify each of the following polynomial expressions as either a sum or a product.

a) $2x + 1$

e) $(x + 5)(x + 2)$

b) $x^2 + 7x + 5$

f) $x(2x + 5)$

c) $(x + 1)(x + 4)$

g) $5x^3 + 8x^2 + 10x + 13$

d) $3x + 7$

h) $2x(x^2 - 3x + 5)$

TR-32. Write an equation to represent the following problem and solve it.

A stick 86 centimeters long is cut into three long pieces (all the same length) and two shorter pieces (both the same length). Each of the shorter pieces is two centimeters shorter than one of the longer pieces. Find the length of a longer piece.

TR-33. Make a table with at least six entries and draw a graph of the equation $y = \sqrt{x} - 2$. Is the graph a line or a curve? What is the smallest x value that can be used?

TR-34.*   From the 12 graphs below, choose one that best fits each of the situations described in parts
           (a) through (d). Copy the graph that fits each description and label the axes clearly with the
           labels shown in the parentheses. If you cannot find a graph you want, sketch your own and
           explain it fully.

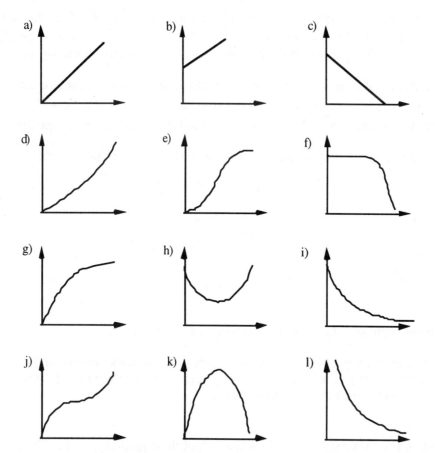

a)   Hulk Hogan held Gorgeous George over his head for a few unsteady moments, and
      then, with a violent crash, he dropped him. (height of Gorgeous / time)

b)   When I started to swim competitively, I initially made very rapid progress, but I have
      found that the better you get, the more difficult it is to improve further. (proficiency /
      amount of practice)

c)   If schoolwork is too easy, you don't learn anything from doing it. On the other hand, if
      it is so difficult that you cannot understand it, again you don't learn. That is why it is so
      important to plan work at the right level of difficulty. (amount learned / difficulty of
      work)

d)   When biking, I try to start off slowly, build up to a comfortable speed, and then
      gradually slow as I near the end of my training. (speed / time)

e)   Make up a story of your own to accompany one of the remaining graphs. In class you
      will give your story to a partner and see whether your partner chooses the same graph
      you wrote the story about.

---

* Adapted from *The Language of Functions and Graphs*, Joint Matriculation Board and the Shell Centre for Mathematical
  Education, University of Nottingham, England.

TR-35.   You have four cards: an ace of spades, an ace of hearts, a two of hearts, and a three of hearts.

    a)   List all possible combinations of <u>two</u> of the cards.

    b)   You draw two cards. What is the probability that both are aces?

    c)   You draw two cards. What is the probability they are both hearts?

    d)   If you draw only <u>one</u> card, what is the probability that it is an ace?

    e)   If you draw two cards, what is the probability that they are <u>both</u> spades?

TR-36.   Solve each of the following equations for y.

    a) $\dfrac{y - 2}{10} = \dfrac{5}{8}$

    b) $\dfrac{y - 2}{10} = \dfrac{x}{6}$

TR-37.   For each statement, find at least one value for n which makes the statement true.

    a) $a^4 \cdot a^n = a^{11}$

    b) $6^n \div 6^4 = 6^{-8}$

    c) $n^n = 1$

    d) $n^2 = 2^n$

TR-38.   The germination rate for zinnia seeds is 78%. This means that on the average 78% of the seeds will sprout and grow. If Jim wants 60 plants for his yard, how many seeds should he plant?

TR-39.   Point A is (1, 1) and point B is (3, 4).

    a)   Describe the slide from A to B.

    b)   Point C is on the line which goes through points A and B. If the x-coordinate of C is 4, what is the y-coordinate? Graph the points and indicate the slide.

TR-40.*   Back to the beakers! The graph below shows how the height of the liquid increases as water is added at a steady rate. Copy each beaker's shape and the graph for x, then plot a graph for beaker A and beaker B.

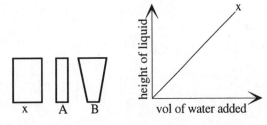

* Adapted from *The Language of Functions and Graphs*, Joint Matriculation Board and the Shell Centre for Mathematical Education, University of Nottingham, England.

5.4 MULTIPLYING POLYNOMIALS AND THE ZERO PRODUCT PROPERTY

TR-41.   If you know the dimensions of a composite rectangle, you can use algebra tiles to build the
          complete rectangle. And once you know what tiles to use to build a composite rectangle, you
          can write its area as a sum.
          Here's an example where the given dimensions are x + 1 and x+ 4:

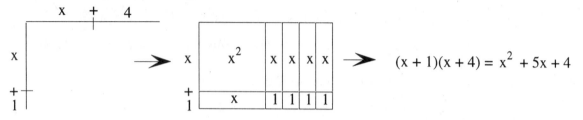

          For each of parts (a) through (f), the dimensions of a composite rectangle are given. Use your
          algebra tiles to construct the indicated rectangle. On your paper, draw the complete composite
          rectangle and then write its area as a product and as a sum.

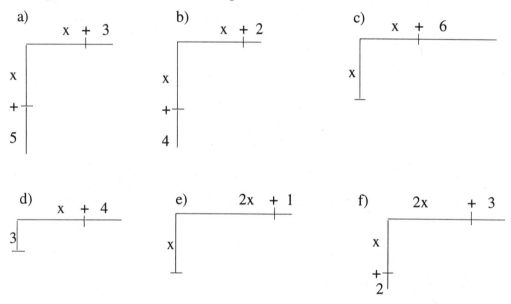

TR-42. We can make our work drawing tiled rectangles easier by not filling in the whole picture. That is, we can show a **generic rectangle** by using an outline instead of drawing in all the dividing lines for the rectangular tiles and unit squares. For example, we can represent the rectangle whose dimensions are $x + 1$ by $x + 2$ with the generic rectangle shown below:

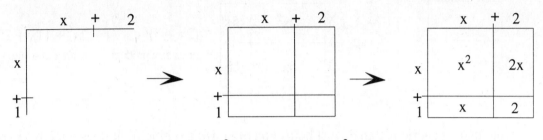

$$(x + 1)(x + 2) = x^2 + 2x + 1x + 2 = x^2 + 3x + 2$$

area as a product   area as a sum

In the last step we found the area of each of the parts of the generic rectangle by multiplying each length and width, and then recorded the area of each part.

Complete each of the following generic rectangles; that is, complete the outline of the parts without drawing in all the dividing lines for the rectangular tiles and small squares. Then, find and record the area of each part. For each completed generic rectangle, write an equation of the form

area as a product = area as a sum.

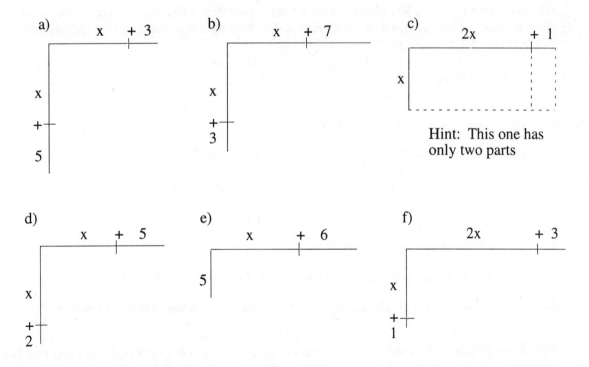

c) Hint: This one has only two parts

TR-43.   You can use generic rectangles to find certain products. For example, to find $(2x + 5)(x + 3)$, you can write in the area of each part, and then add:

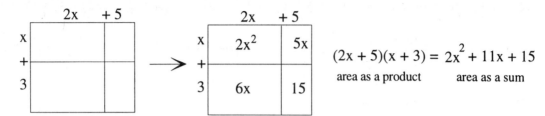

$$(2x + 5)(x + 3) = 2x^2 + 11x + 15$$
area as a product   area as a sum

Note that a generic rectangle just helps you organize a problem. It does not have to be drawn accurately to scale. In the example above, you can see that the part representing $5x$ is probably not as big as it would really be compared to the part representing $6x$.

Find each of the following products by drawing and labeling a generic rectangle. In some of the problems, the generic rectangles will have only two parts.

a) $(x + 5)(x + 4)$ d) $2x(x + 3)$

b) $(x + 2)(x + 8)$
                                           e) $3(2x + 4)$

c) $x(x + 10)$ f) $(x + 30)(x + 20)$

TR-44.   **Finding the Area of a Rectangle** You've been practicing *finding the area of a rectangle given its dimensions*, or if we say it algebraically, writing the product of two factors as a polynomial. An example of this description is
$$(x + 2)(x + 8) = x^2 + 10x + 16.$$
Write another example.

TR-45.   a)   Find each of the following products:

    $(5.75)(8.33)(0)(4.724)$ $(3.27 \cdot 10^{-2})(0)$

    $(5.75)(0)(8.33)(4.724)$ $(0)(x)(0)$

    $(6^3)(x)(0)(y)(4^2 - 1)$ $(0)(x^2)$

   b)   If $(35)(x)(4)(7)(29) = 0$, what must be the value of x ?

   c)   If $A \cdot B = 0$, what can you say about the possible values of A and B ?

   d)   If the product of several numbers is zero, what can you say about the numbers?

TR-46.   **The Zero Product Property** In the preceding problem you examined the **zero product property:** If the product of numbers is zero, then at least one of the numbers must be zero.

Now consider the equation $(x + 3)(x - 1) = 0$. Use Guess and Check to determine a value of x that will make the equation a true statement. Is there more than one value for x that makes the equation true? By substituting the values for x, demonstrate how each of your solutions works.

TR-47.   The Smith family had triplets, and exactly two years later they had twins. Several years after that the total of the ages of the five Smith children was 86 years. How old were the triplets at that time?

TR-48.   Look back at problems TR-12, TR-32, and TR-47 and compare the equations you wrote in each case. Write one or two sentences about what you notice.

TR-49.   Make a table with at least eight entries and draw a graph of the rule $y = 2^x - 1$.

TR-50.   Sketch a graph to show how human height varies with age.

TR-51.   Solve each of the following equations.

a) $3(5x - 7) = 4(4x + 8)$

b) $9x - 875 = 4x + 1200$

c) $\dfrac{5}{x + 2} = \dfrac{3}{x}$

d) $2(w + 4) - 6(w - 7) = 50$

TR-52.   Start at point A $(-1, 2)$ and slide to point B $(3, 5)$. Find the coordinates of a point C so that A, B, and C are on the same line.

TR-53.   Compute the numeric value of each of the following expressions.

a) $-2 + 5 \cdot 6 \div 3 - 1$

b) $(-2)^5 - 6 \div 3 - 1$

TR-54.   Deli sandwiches can be bought at Giuseppi's in a variety of lengths. No matter how long a sandwich is, it is uniformly made, so a short version of a sandwich differs from a longer version only in length. The shortest version of a *Sub* is 6 inches long, weighs $\frac{3}{4}$ pounds, and costs \$2.35. The longest version of the *Sub*, the *Party Giant*, is 21 feet long.

a)   How much does the *Party Giant* weigh?

b)   Giuseppi charges by the weight of the sandwich. How much does the *Party Giant* cost?

TR-55.   Solve each of the following equations for x. That is, what number(s) make each equation true?

a) $0x = 5$

b) $7x = 7x$

5.5 FACTORING RECTANGLES

TR-56.   What if we knew the area of a rectangle and we wanted to find the dimensions? We would
have to work backwards. Let's start with the area represented by $x^2 + 5x + 6$. Normally, we
would not be sure whether the polynomial represents the area of a rectangle, but we could find
out by using algebra tiles to try to form a rectangle. We tried this with $x^2 + 5x + 6$ before, and
it worked:

We can see that this rectangle with area $x^2 + 5x + 6$ has dimensions $x + 2$ and $x + 3$.

Use algebra tiles to build rectangles with each of the following areas. Draw the composite
rectangle and write its dimensions algebraically as in the example above. Check your work
with your group.

a) $x^2 + 6x + 8$

b) $x^2 + 5x + 4$

c) $2x^2 + 8x$

d) $2x^2 + 5x + 3$

TR-57. Suppose that each of the generic rectangles shown below represents a composite rectangle made with algebra tiles. (The parts are not necessarily drawn to scale.) Copy each composite rectangle and find its dimensions. For example,

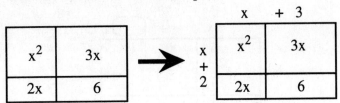

a)

x^2	5x
3x	15

d)

2x | $2x^2$ | 10x |

b)

x^2	4x
3x	12

e)

x^2	5x
2x	10

c)

x^2	6x
3x	18

f)

x^2	4x
4x	16

TR-58. Copy and solve each of the following Diamond Problems.

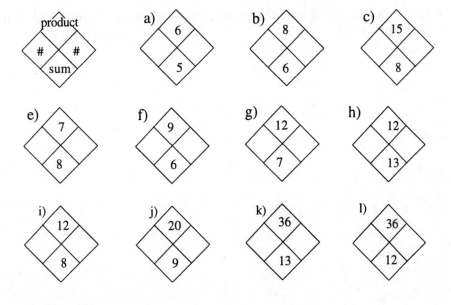

TR-59.   Let's go back to the problem of finding the dimensions of a rectangle when you know its area. You can try to replace the tiles with a generic rectangle and work backwards. For example, start with $x^2 + 8x + 12$, then draw a generic rectangle and fill in the parts you know:

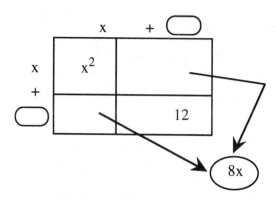

Now, you know the sum of the areas of the two unlabeled parts must be $8x$. But how should the $8x$ be split between the two parts? The $8x$ could be split into $7x + 1x$, or $6x + 2x$, or $3x + 5x$, or $4x + 4x$. There is additional information that will help you decide which split is correct: the numbers that go in the two ovals must have a product of 12.

a)   Could a Diamond Problem help you decide how should the $8x$ be split? If so, which Diamond Problem?

b)   Complete the generic rectangle and label its dimensions.

TR-60.   Use the "working backwards" method of problem TR-59 and the Diamond Problems from TR-58 to find the **factors** of each of the following quadratics. The factors are the binomials which represent the length and width of the rectangle. Draw a picture for each problem and write an algebraic equation of the form *area = (length)(width)*.

a) $x^2 + 7x + 10$ (It's easier to think of the factors of 10 first and then find the pair whose sum is 7.)

b) $x^2 + 6x + 8$

c) $x^2 + 8x + 15$

d) $x^2 + 6x + 9$

e) $x^2 + 9x + 20$

TR-61. You've practiced factoring quadratic polynomials by drawing generic rectangles with four parts. Similarly, you can use generic rectangles with two parts to factor binomials (polynomials with two terms). For example, we could factor $2x^2 + 10x$ as $2x(x + 5)$ by drawing the following generic rectangle:

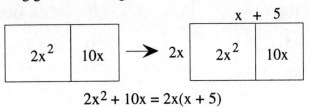

$$2x^2 + 10x = 2x(x + 5)$$

Or, we could factor $2x^2 + 10x$ as $x(2x + 10)$ by drawing this generic rectangle:

x + 5

| $2x^2$ | $10x$ |

2x | $2x^2$ | $10x$ |

$$2x^2 + 10x = 2x(x + 5)$$

Factor each of the following binomials; in other words, find the dimensions of a generic rectangle with each given area. Each generic rectangle in this problem has only two parts. Draw and label each generic rectangle as shown in the examples above.

a) $x^2 + 7x$

b) $3x^2 + 6x$

c) $3x + 6$

d)   There is more than one way to factor $3x^2 + 6x$. Compare your answer in part (b) to those of your group mates and write down all the solutions your group found.

TR-62.   Use a generic rectangle to find each of the following products.

a) $(x + 10)(x + 3)$

b) $(x + 7)(x + 9)$

c) $(2x + 11)(3x + 5)$

TR-63.   Find each of the following products. Use the same method you used in problem TR-62 -- just apply what you know about positive and negative numbers.

a) $(x - 8)(x + 5)$ e) $(x + 3)(x - 3)$

b) $(x - 7)(x - 10)$ f) $(5x + 4)(5x - 4)$

c) $(3x - 2)(x + 5)$ g) $(x + 3)(x + 2)$

d) $(x - 8)(x - 8)$

TR-64.   Use algebra tiles to explain why ...

a) $2x^2$ does not usually equal $2x$.

b) $(3x)(3x) = 9x^2$.

TR-65.   Leilani drew a triangle in which the measure of the second angle was eight degrees more than the measure of twice the first, and the measure of the third angle was 12° less than the measure of the first. Sketch and label the triangle, then write an equation and solve it to find the measure of each angle. [Recall that for every triangle the sum of the measures of the angles is 180 degrees.]

TR-66.   Start with a certain number, add two thirds of it, then add half of it, and then add a seventh of it. When all the numbers are added together, the sum is 97. What is the starting number? (This problem appears in the Rhind Papyrus, which is probably the oldest mathematical book known, written around 1650 BC. in Egypt.)

TR-67.   There is something peculiar about each of the following equations. Try to solve them in the usual way and explain what happens.

a) $3(9 - 2d) + 2(2d + 8) = 6 - 2d$

b) $s + 2(s + 1) + 3(s + 2) = 6s + 8$

TR-68.   Show how to answer each of the following percent problems by setting up an equation and solving it.

a)   If Percy has earned 237 points out of a possible 280 on all tests so far, what percent of the total points has he earned?

b)   If 92% of the 280 points were required for an A, how many points would Percy need to get to earn an A ?

c)   Two hundred-thirty seven is 92% of what number?

d)   Percy has just one more test to take, and it's worth 100 points. What is the lowest score Percy could get on the test and still get 80% of the total points, and so earn a B ?

TR-69. Triangle ABC, ΔDEF, and ΔGHI are similar right triangles. Find each missing side by writing an equation and then solving it.

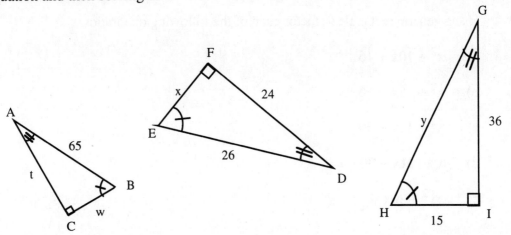

TR-70. Solve each of the following equations for y.

a) $2y + 8 = 6$

b) $2y + 3x = 6$

TR-71. Find the value of each of the following expressions if $x = -2$, $y = 1$, and $z = 3$.

a) $3x^2 - 2(y - 3)$

b) $(z + y \div x)^3$

TR-72. The survival rate for starfish is remarkably low. If it weren't, then life would certainly be very different here on earth. A female starfish lays 1,000,000 eggs at a time. If all of the eggs hatched and 50% were female baby starfish, how many starfish would be born in ...

a) in the second generation?

b) in the third generation?

c) in the fourth generation?

TR-73. Beakers, continued! The graph below shows how the height of the liquid increases as water is added at a steady rate. Copy beaker x and the graphs for beakers x, A, and B, then sketch the shapes of beaker A and beaker B.

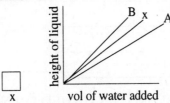

5.6 FACTORING POLYNOMIALS

TR-74.   Use a generic rectangle to factor each of the following quadratics.

   a) $x^2 + 10x + 16$

   b) $x^2 + 15x + 56$

   c) $x^2 + 11x + 30$

   d) $x^2 - 11x + 30$

   e) $5x^2 + 30x$

TR-75.   **Sums and Products**

   a)   Use your Sums and Products resource page to record integer pairs whose sums and products are indicated in the table

Sum	Product	Integer Pairs
5	6	2, 3 or 3, 2
7	6	
13	12	
10	16	
6	12	
8	16	
15	56	
11	30	
5	-6	
2	-35	
1	-72	
3	-10	
10	-24	
0	-16	
23	-24	
0	-32	

Sum	Product	Integer Pairs
-5	6	
-7	6	
-11	30	
-8	16	
-10	16	
0	16	
-10	24	
-9	18	
-5	-6	
-2	-35	
-3	-28	
-1	-12	
-10	-25	
0	-25	
-1	-42	
-12	-13	

[PROBLEM CONTINUED ON NEXT PAGE]

TR-75. continued

b)   Judy says the entries in the sum-product table in part (a) are just Diamond Problems in disguise. Explain what she means.

c)   Look again at problem TR-74, parts (a), (b), (c), and (d). Which line in the Sums and Products table is related to each of them? Circle the related lines in your table.

d)   We can use a line in the sum-product table to write a quadratic polynomial and its factors. For example, $x^2 - 10x + 24 = (x - 4)(x - 6)$. Give three more examples.

e)   Use the table to write a quadratic polynomial that has <u>no</u> factors.

TR-76.   Use your experience with sums and products of integers to help you factor each of the following quadratics.

a) $x^2 - 6x + 8$

b) $x^2 - 9x + 18$

c) $x^2 + 4x - 21$

d) $x^2 - 3x - 10$

e) $x^2 - x - 20$

TR-77.   Factor each of the following quadratics. You can think of $x^2 - 49$ as $x^2 + 0x - 49$.

a) $x^2 - 49$

b) $x^2 - 16$

c) $x^2 - 25$

d)   An expression such as those above is called a "difference of two squares." Why?

e)   Write another quadratic which is the difference of two squares, then factor it.

f)   Another expression that is a difference of two squares is $16x^2 - 9$. Factor it.

TR-78.   Now factor these quadratics. Explain what is special about them.

   a) $x^2 - 10x + 25$

   b) $x^2 + 20x + 100$

   c) $x^2 - 12x + 36$

   d)   Parts (a), (b), and (c) are called "perfect square trinomials." Why?

   e)   Give another example of a perfect square trinomial. Factor it.

   f)   The polynomial $9x^2 + 30x + 25$ is also a perfect square trinomial. Find its factors.

TR-79.   Find each of the following products.

   a) $(x + 13)(x + 3)$ g) $7x(3x + 5)$

   b) $(x - 8)(x + 10)$ h) $-2x(4x + 11)$

   c) $(x - 9)(x - 8)$ i) $(2x - 5)^2$

   d) $(x - 9)(x + 9)$ j) $(3x + 7)^2$

   e) $(2x + 7)(3x)$ k) $(x + 11)^2$

   f) $4(5x - 2)$

TR-80.   Factor each of the following quadratics.

   a) $x^2 - 6x + 5$ f) $6x^2 - 12x$

   b) $x^2 - x - 6$ g) $x^2 + 9x + 7$

   c) $x^2 - x - 42$ h) $x^2 - 3x - 88$

   d) $6x + x^2 - 16$ i) $x^2 - 25$

   e) $x^2 - 100x + 2500$

TR-81.   Factor each of the following quadratics. Work together and treat this set as a puzzle. If the
           numbers you try don't work for one expression, they may work for another.

   a) $x^2 - 2x - 24$ e) $x^2 - 10x - 24$

   b) $x^2 + 11x + 24$ f) $x^2 + 8x + 24$

   c) $x^2 - 10x + 24$ g) $x^2 - 23x - 24$

   d) $x^2 + 5x - 24$ h) $x^2 + 25x + 24$

TR-82.   Show that $x = 2$ and $x = -3$ are solutions to the equation $(x - 2)(x + 3) = 0$.
(Hint: Substitute $x = 2$ for both x's and show the product is zero. What happens if you do the same thing with $x = -3$?)

TR-83.   a)   If $(x + 2)(x - 3) = 0$, what must be true about either the factor $x + 2$ or the factor $x - 3$? For a hint, look back at TR-46.

b)   Use what you wrote in part (a) to find two values for x that make equation $(x + 2)(x - 3) = 0$ true.

c)   Solve the equation $(x - 2)(x + 5) = 0$

d)   Solve the equation $(x + 3)(2x - 3) = 0$

TR-84.   Use the factorizations you found in problem TR-81 to help you solve these equations:

a) $x^2 - 2x - 24 = 0$

b) $x^2 + 11x + 24 = 0$

c) $x^2 - 10x + 24 = 0$

d) $x^2 + 5x = 24$

TR-85.   We can use the idea of a generic rectangle to find different kinds of products. For each of the following products, draw a generic rectangle and label the dimensions. Then find the given product by finding the area of the rectangle. The first answer is given so you can check your method.

a) $(x + 3)(x^2 + 4x + 7)$ Answer: $(x + 3)(x^2 + 4x + 7) = x^3 + 7x^2 + 19x + 21$

b) $(3x - 2)(4x^2 + 2x + 1)$

c) $(2x + 1)(x^2 - 3x + 4)$

d) $(x - 2)(x^2 + 2x + 4)$

TR-86.   Some expressions can be factored more than once! For example, suppose you want to factor
          $3x^3 - 6x^2 - 45x$ as completely as possible. Using a generic rectangle, you could factor
          $3x^3 - 6x^2 - 45x$ as $(3x)(x^2 - 2x - 15)$:

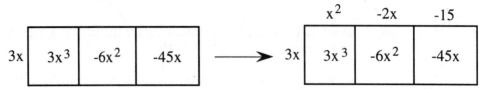

However, $x^2 - 2x - 15$ can also be factored, namely $x^2 - 2x - 15 = (x + 3)(x - 5)$.
                                        Check this!
Thus, $3x^3 - 6x^2 - 45x$ can be factored as the product $3x(x + 3)(x - 5)$.
                         How can you tell if you are done?

Factor each of the following polynomials as completely as possible.

a) $5x^3 + 15x^2 - 20x$

b) $3x^2y - 9xy - 30y$

c) $7x^2 - 847$

TR-87.   Factor each of the following quadratics. These will be more challenging because the coefficient
          of x^2 is not 1 as in previous problems. You may need to use Guess and Check.

a) $3x^2 + x - 2$

b) $3x^2 + 7x + 2$

c) $2x^2 - 3x - 5$

d) $2x^2 - 5x - 3$

e) $5x^2 - 13x + 6$

TR-88.   Use four 2's, two sets of parentheses, and any operations you wish to write an expression
          which has the greatest value.

For each of problems TR-89 and TR-90, write an equation and solve it. Be sure to identify what the
variable represents if you do not make a Guess and Check Table.

TR-89.   Two cars, an Edsel and a Studebaker, are 635 kilometers apart. They start at the same time
          and are driven toward each other. The Edsel travels at a rate of 70 kilometers per hour and the
          Studebaker travels at 57 kilometers per hour. In how many hours will the two cars meet?

TR-90.   Mr. Nguyen is dividing $775 among his three daughters. If the oldest gets twice as much as the youngest, and the middle daughter gets $35 more than the youngest, how much does each girl get?

TR-91.   Sketch a graph and label the axes for each of the following descriptions.

   a)   As the temperature increases, the volume of the material increases at a steady rate.

   b)   As the temperature increases, the number of people at the beach increases until the temperature reaches 110°, then the number levels off.

TR-92.   Find y.

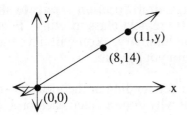

TR-93.   Use a generic rectangle to find each of the following products. (You may want to look back at the example for TR-43.)

   a) $(x + 1)(x + 1)$ b) $(x + 5)(x + 2)$

   c) $2x(x + 5)$ d) $(2x + 1)(x + 5)$

   e) $(x + y)(x + y)$

TR-94.   **Chapter 5 Summary: Rough Draft** The main ideas of this chapter are:

1.   The total area of a rectangular figure equals the sum of the areas of the individual parts.

2.   The area of a rectangle can be expressed in two equivalent ways:

"area as a product = area as a sum"

• We can always use multiplication to change an expression from a product to a sum.

• We can sometimes use factoring to change an expression from a sum to a product.

3.   Polynomials can be added or subtracted.

Write your answers to the following questions in rough draft form **on separate sheets of paper**, and be ready to discuss them with your group at the next class meeting. Focus on the **content**, not neatness or appearance, as you write your summary. You will have the chance to revise your work after discussing the rough draft with your group.

a)   Select four problems which best illustrate your understanding of the ideas listed above. Show all your work and complete solutions. Tell why you selected each problem.

b)   Select a problem you didn't understand before but now know how to do. Show all your work and a complete solution. Explain why you chose the problem.

5.7 SUMMARY AND REVIEW

TR-95. **Chapter 5 Summary: Group Discussion** Take out the rough draft summary you completed in TR-94. Use this time to discuss your work and use homework time to revise your summaries as needed.

Take turns to describe the problems group members chose to illustrate the main ideas of the chapter. Each group member should:

- explain the problems he or she chose to illustrate the main ideas;

- explain why he or she chose those particular problems; and

- explain a problem he or she didn't understand before, but now can solve, and explain why the problem was chosen.

This is your chance to make sure your summary is complete, update your Tool Kits, and work together on problems you may not be able to solve yet.

TR-96. Complete the following generic rectangles so that the length, width, and area of each part are labeled. Write the area of each rectangle as a product and as a sum.

a)

$$x + 5$$

x	–	–
+		
3	–	–

b)

| x | $4x$ |
| -2 | ___ |

c)

| – | x^2 | $-4x$ |
| – | $+5x$ | -20 |

d)

| ___ | $15x^2$ | $+ 10x$ | $- 5$ |

TR-97. Use generic rectangles to factor these quadratics:

a) $x^2 + 9x + 14$

b) $x^2 + 8x + 7$

c) $x^2 - 5x + 4$

d) $x^2 + 2x - 15$

e) $x^2 - 5x - 24$

f) $x^2 + 10x - 24$

TR-98.   Find each of the following products.

   a) $3x(2x - 6)$ b) $(x - 9)(x + 7)$

   c) $(3x - 1)(x - 5)$ d) $3x(2x^3 - x + 4)$

TR-99.   Rewrite each of the following sums (or differences) of polynomials in simpler form.

   a) $(5x^2 + x - 1) + (3x^2 - 4x - 8)$

   b) $(6x^2 - 8x + 7) - (2x^2 + 6x - 7)$

   c) $(8x^2 - 8x + 10) + (2x^2 + 8x - 6)$

   d) $(x^2 - x - 1) - (2x^2 + x - 1)$

TR-100.   Scott wants to enlarge each side of this rectangle by a factor of 3 :

   Draw a picture of the enlarged rectangle and label its dimensions. Then, write
   algebraic expressions for the perimeter and the area of the new, enlarged rectangle.

TR-101.   Anita drew this diagram of a frame for an x by x + 1 rectangular picture:

$$2x - 1$$
$$x + 1$$
$$2x + 5 \quad x$$

   Write an algebraic expression to represent the total area of the frame.

TR-102.   **Chapter 5 Summary: Revision** Use the ideas your group discussed for TR-95 to revise
   the rough draft of a Chapter 5 Summary you wrote for TR-94. Your presentation should be
   thorough and organized, and should be done on a separate piece of paper.

Algebra Tiles

Glue this page to a piece of lightweight cardboard (an empty cereal box works) or another piece of paper. Carefully cut out the individual rectangles and the small and large squares. Keep them in a re-sealable envelope or plastic bag. Store the bag of tiles in your notebook.

Centimeter Grid Graph Paper

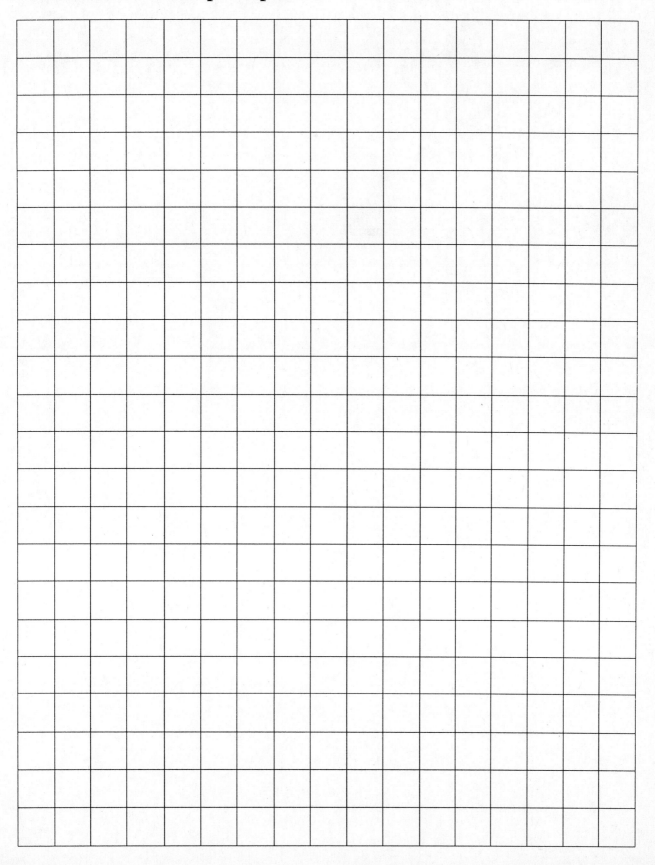

Centimeter Grid Graph Paper

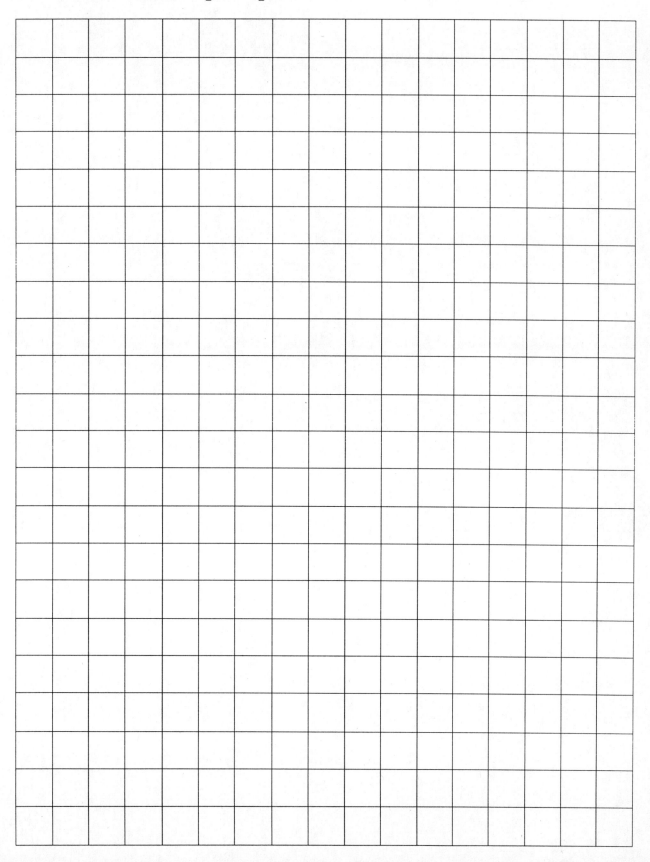

Algebra Tool Kit: the "what to do when you don't remember what to do" kit

Algebra Tool Kit: the "what to do when you don't remember what to do" kit

TR-11. Composite Rectangles and Area

Number of x^2 tiles	Number of 1x tiles	Number of 1 tiles	Is a composite rectangle possible?	Sketch	Algebraic Expression for the Area of the Composite Rectangle
1	3	2	Yes		$x^2 + 3x + 2$
1	5	3			
1	4	4			
1	6	5			
1	3	9			
1	4	3			
1	7	10			

‡TR-75. Sums And Products

a) For each sum-product pair, find two integers whose sum is the number in the left-hand column and whose product is the number in the right-hand column. For example,

Find two integers whose sum is 5 and whose product is 6.
Solution: The integers 2 and 3 work, since $2 + 3 = 5$ and $2 \cdot 3 = 6$.

Note: There may be some sum-product pairs for which no pair of integers works.

Sum	Product	Integer Pairs
5	6	2, 3 or 3, 2
7	6	
13	12	
10	16	
6	12	
8	16	
15	56	
11	30	
5	-6	
2	-35	
1	-72	
3	-10	
10	-24	
0	-16	
23	-24	
0	-32	

Sum	Product	Integer Pairs
-5	6	
-7	6	
-11	30	
-8	16	
-10	16	
0	16	
-10	24	
-9	18	
-5	-6	
-2	-35	
-3	-28	
-1	-12	
-10	-25	
0	-25	
-1	-42	
-12	-13	

Chapter 6
THE AMUSEMENT PARK: Graphing and Systems of Equations

Chapter 6

AT THE AMUSEMENT PARK:
Graphing and Systems of Linear Equations

THE AMUSEMENT PARK

So far in this course you've graphed a lot of equations and focused a lot of attention on solving word problems by writing and solving equations. In this chapter you'll extend your skills for solving word problems as you examine problems which can be modeled by pairs of related equations. In solving these "systems of equations" you'll see how the algebraic approach to solving equations is related to the graphical approach. You'll apply these two approaches as you develop ways to solve problems like the one presented by the Amusement Park problem:

> Jim and his friends are going to the amusement park and find that they have two ticket options. In one option each person could buy an admission ticket for $25.50 and then pay 85¢ for each ride. The other option is to buy an admission ticket for $11.90 and then pay $2.55 a ride. What do you think Jim should do?

In this chapter, you will have the opportunity to:

- interpret graphs of data;

- develop a relatively quick method for graphing a linear equation by using its y-form and two points;

- use the intersection of two graphs to solve problems involving two sets of information which lead to two equations;

- solve a system of two equations algebraically by the substitution method; and

- find the intersection of the line $y = 0$ and the parabola $y = ax^2 + bx + c$ by solving the equation $ax^2 + bx + c = 0$ with the use of factoring and the zero product property.

Many of the problems in this chapter require graph paper. Be sure you have an adequate supply.

CHAPTER CONTENTS

6.1   Using Linear Equations to Model Problems

AP-1.   The width (in feet) of a rectangle is three more than a secret number and the length of the rectangle is five times the width.

     a)   Draw a diagram of a rectangle. Choose a letter as a variable to represent the secret number and write an algebraic expression for the width of the rectangle. Label your diagram.

     b)   Write an algebraic expression for the length of the rectangle using the variable you chose in part (a). Write the expression on your diagram.

     c)   Use your expressions from parts (a) and (b) to write an equation that says the perimeter of the rectangle is 144 feet. (Remember that the perimeter of a figure is the total distance around its border.)

     d)   Use your equation from part (c) to find the secret number. What are the length and width of the rectangle?

     e)   What is the area of the rectangle?

AP-2.   When Ellen started with Regina's favorite number and tripled it, the result was twelve more than twice the favorite number. Define a variable, then write an equation and use it to find Regina's number.

AP-3.   I'm thinking of a number. If I add three to the number, multiply the sum by two, and subtract five, the result will be four more than three times my original number. What is my number? Define a variable, then write an equation and solve it.

AP-4.   **Grading Quiz #10** Dick and Crystal are Ms. Speedi's teaching assistants. Four minutes after Dick started to grade Quiz #10, Crystal joined him. Dick graded at a rate of three answers per minute, while Crystal graded at the rate of five answers per minute. They continued grading at these rates until they were finished. The diagram below shows Dick's and Crystal's contributions to grading Quiz #10. The whole segment represents a completed job.

               Dick   Crystal & Dick

     a)   Let x represent the number of minutes Dick spent grading quizzes. For each of the following descriptions, write an expression using x : .

        (1)   the number of minutes Crystal graded quizzes

        (2)   the number of answers Dick checked

        (3)   the number of answers Crystal checked

        (4)   the total number of answers checked by both Dick and Crystal

[PROBLEM CONTINUED ON NEXT PAGE]

AP-4. continued

  b)   After five minutes, how many answers had Dick checked? How many answers had
       Crystal checked?

  c)   After 10 minutes, how many answers had Dick checked? How many answers had
       Crystal checked?

  d)   Write an equation that states that Dick and Crystal checked a total of 36 answers. Solve
       the equation to find out how long Dick had been grading when 36 answers had been
       checked. (Hint: Use the pattern in parts (b) and (c) to help you write the equation.)

  e)   How long did Crystal work if she and Dick checked a total of 220 answers?

AP-5.   **The Bike Race** Cindy and Dean are engaged in a friendly 60-mile bike race that started at
         noon. The graph below represents their race. Notice that Dean, who doesn't feel a great need
         to hurry, stopped to rest for a while.

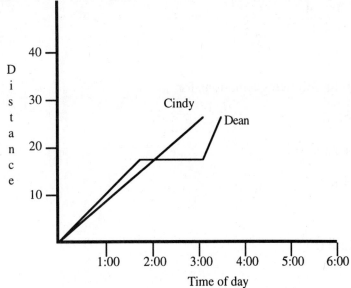

To answer each of the following questions, show how to use the graph, and then write your
responses. Do this on your copy of the AP-5 resource page.

  a)   At what time (approximately) did Cindy pass Dean?

  b)   About how far has Cindy traveled when she passes Dean?

  c)   You can see that Dean will catch up to Cindy again if they both continue at their same
       rates. At approximately what time will this happen? About how far have they both
       traveled to this point?

  d)   Does Dean travel faster or slower after his rest?

  e)   About how long did Dean rest?

  f)   If Cindy continues at a steady pace, about how long will it take her to complete the race?

AP-6.   It is often useful to know whether a particular point lies on a line without actually graphing the line.

a)   Graph the equation $y = 3x + 2$ for $x = -3, -2, \ldots , 2, 3$.

b)   On your graph, clearly mark the points $(2, 6), (1, 5), (-2, -4), (0, 1)$ and $(-1, -1)$.

c)   Does the point $(1, 5)$ make the equation $y = 3x + 2$ true or false? Explain how you know.

d)   Now check whether each of the points $(2, 6), (-2, -4), (0, 1)$ and $(-1, -1)$ makes the equation $y = 3x + 2$ true or false.

e)   Look for a relationship between the points from parts (b) and (c) that lie on (or off) the graph and those that make the equation true (or false). Describe what you observe.

AP-7.   Without drawing a graph, determine which, if any, of the listed points are on the graph of the equation $y = 3x - 4$.

$(-1, -4)$ $(2, 2)$ $(1, -1)$ $(3, 3)$

AP-8.   Solve each of the following equations for x.

a) $2x + 23 = 0$

b) $2x - 9 + 3x = 16$

c) $3x + 8 = k$

d) $x + 15 = t$

e) $x(w + 2) = 0$

AP-9.   For each of the following equations, make a table with the indicated values for x and then graph the equation.

a) $y = -2x + 3$ for $x = -4, -3, \ldots , 4, 5$

b) $y = \frac{1}{2}x - 4$ for $x = -4, -2, 0, \ldots , 10, 12$

AP-10.   Solve each of the following equations for x. Show all steps.

a) $6x + 7 = 8x - 14$

b) $6x - 2(x + 8) = 24$

c) $\frac{5 - x}{7} = 3$

d) $\frac{10}{x - 6} = \frac{15}{x - 3}$

e) $-3(2x + 5) = 87$

f) $\frac{x}{3} + 6 = -45$

g) $17(x - 3) = 0$

h) $(x + 1)(x - 3) = 0$

AP-11.   Solve each of the following equations for y. Show each step.

a) $2x + y = 8$

b) $2x - y = 8$

c) $3x + 2y = 6$ (Check your result: $y = \frac{6 - 3x}{2}$ or $y = 3 - \frac{3x}{2}$)

d) $3x - 2y = 6$

e) $6x + 2y - 14 = 0$

AP-12.   Without drawing a graph, determine which, if any, of the listed points are on the graph of the equation $y = -2x - 3$.

$(-1, -1)$ $(-2, -2)$ $(3, -7)$ $(4, -11)$

AP-13.   Find the area of the shaded region of this rectangle:

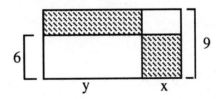

AP-14.   Rewrite each of the following expressions by multiplying the binomials in parts (a) and (b) and factoring the polynomials in parts (c) through (f).

a) $(x - 5)(x + 7)$

b) $(x + 10)(x - 10)$

c) $x^2 + 10x + 16$

d) $x^2 + 0x - 25$

e) $x^2 - 6x + 8$

f) $5x^2 + 15x$ Look for common factors.

AP-15.   Julio runs $\frac{3}{10}$ mile in $1\frac{1}{2}$ minutes. If he keeps running at that rate, how long will it take him to run a whole mile? Show how you know.

AP-16.   José saved $12.60 when he bought his new sneakers on sale for 30% off of the original price. Write and solve an equation to find how much the sneakers cost originally.

AP-17.   Write and solve an equation using ratios to answer each of the following questions.

a)   Sixteen is what percent of 25 ?

b)   Sixteen percent of 25 is what?

c)   Twenty-five is what percent of 10 ?

AP-18.   Write your own "thinking of a number" problem like AP-3 that will give you a complicated linear equation. Copy your problem on a separate piece of paper, or an index card, so that you can share it with your group tomorrow.

6.2 Graphing Linear Equations using Two Points

> In the remainder of this chapter many of the graphs will require a full sheet of graph paper so that the graphs you make are clear, accurate, and large enough for you to analyze. Use a straightedge or ruler to draw the graphs of lines.

AP-19.   So far in this course you have graphed equations by first selecting several different values for x and then calculating the corresponding y-values. Each of the equations you've graphed has been written in **y-form**; that is, y is alone on the left-hand side of the equation while x and numbers are on the right. Problems AP-19 through AP-24 will help you develop a faster method for graphing linear equations.

a)   Use three different x-values to find three points that lie on the graph the equation
$$y = 2x - 5.$$
The members of your group should all choose different x-values.

b)   Draw and label coordinate axes on a full sheet of graph paper. Scale the axes so that each tick mark is one unit. Use the three points you found in part (a) to graph the equation $y = 2x - 5$. Label the points you plotted with their coordinates and label the line with its equation.

b)   Now check with your group: Does the graph of equation $y = 2x - 5$ depend on which x-values you selected or are your graphs of the equation $y = 2x - 5$. the same? (Remember that lines go on forever.)

AP-20.   Here are two more equations in y-form:

$$(1) \quad y = 2x + 3 \qquad\qquad (2) \quad y = 2x - 1$$

a)   Carefully graph each equation on the set of axes you used in AP-19. Use –2, 0, and 2 as x-values for each equation. Label each graph with its equation.

b)   Recall that the x-coordinate of any point on the y-axis is zero. Thus any point on the y-axis can be written in the form (0, y).

Examine each of your graphs to find **the y-coordinate of the point where the line crosses the y-axis**. This point is called the **y-intercept** of the line. Its coordinates are of the form **(0, y)**. For example, the line $y = 2x - 5$ that you graphed in AP-19 has y-intercept (0, –5).

Add information about the y-intercept of a line to your Tool Kit.

c)   Now, for each line you graphed in part (a) copy and complete the following sentence:
"The y-intercept of the line _____ is _____ ." .

d)   Compare the y-intercept for each line with the constant (number) on the far right of the y-form of the line's equation. Write a sentence to describe what you observe.

AP-21.   Here are some linear equations in y-form. Use your observations in the previous problem to
          find the y-intercept of the graph of each equation. Write your answers in the form

                    "The y-intercept of the line _____ is _____ ."

          a) $y = 2x + 3$

          b) $y = -x - 1$

          c) $y = 2x - 4$

          d) $y = -3x + 2$

AP-22.   In problem AP-19 each member of your group graphed the equation $y = 2x - 5$ using three x-
          values. Although you didn't all use the same x-values, your graphs were the same line.

          a)   Assume that you already know the coordinates of one point that lies on the graph of a
                linear equation. What is the least number of additional points you need to plot in order to
                sketch a graph of the equation?

          b)   Use your observations from problem AP-20 to find one point on the graph of $y = 2x - 5$.
                You don't need to do any computation, but if you must substitute, use $x = 0$.

          c)   Now choose **any x-value other than zero** and find its corresponding y-value.

          d)   Plot the two points you found in parts (b) and (c) on a pair of coordinate axes drawn on a
                full sheet of graph paper and scaled so that each tick mark is one unit. Carefully use a
                ruler to draw a line through the two points and label the line with its equation.

          e)   Pick another point on the line and check to see if it makes the equation $y = 2x - 5$ true.

          f)   Compare your graph from part (d) to your graph in problem AP-19. Write one or two
                sentences to describe what you observe.

AP-23.   In problem AP-22 you graphed a line by using the y-form of an equation to find two points on
          the line -- its y-intercept and one other point. To get an accurate graph using this method, it
          helps if the points you choose are not too close together. It is also a good idea to always check
          the accuracy of your graph by finding a third point and making sure it lies on the line formed
          by the first two points.

          Use the handy **two-point y-form method** to graph the four equations in problem AP-21
          on the same set of axes you used in AP-22. Be sure to label each line with its equation and to
          check the accuracy of your graphs.

AP-24. Although it's often easy to graph equations that are written in y-form, not all equations are. For example, the equation

$$x + y = -1$$

is not in y-form. However, you could rewrite it in y-form by solving it for y as shown in the following example.

  a)   Copy this example in your Tool Kit:

      To rewrite the equation $x + y = -1$ in y-form, you just
      need to solve it for y. You can do this by subtracting x
      from each side of the equation:

$$\begin{array}{rcr} x + y &=& -1 \\ -x & & -x \\ \hline y &=& -1 - x \end{array}$$

      Note that you could also write $y = -1 - x$ as $y = -x - 1$.

  b)   Write the equation $3x + 2y = 2$ in y-form by solving it for y. Show each step.

AP-25. Write each of the following equations in y-form.

  a) $3x + y = -4$

  b) $y - 4x = 8$

  c) $2x - y = 3$ Be sure you notice that it is $-y$.

  d) $6x + 2y = 10$

AP-26. Graph each of the following linear equations using the two-point y-form method. You will need to change some of the equations into y-form first.

  a) $y = -2x + 6$ d) $2x - y = 5$

  b) $y = 3x - 3$ e) $6x + 2y = 7$

  c) $4x + y = 1$ f) $3x - 6y = -24$

AP-27. Without drawing a graph, determine which, if any, of the following points are on the graph of $y = -3x - 5$: $(0, -5)$, $(2, -11)$, and $(-3, -14)$.

AP-28. Can each of these equations be graphed with the two-point y-form method? Why or why not? You do not need to graph the equations.

  a) $y = 2x + 1$

  b) $y = 2x$

  c) $y = x^2$

AP-29.   Arrange the tiles represented by $2x^2$, $9x$ and 9 into a composite rectangle.

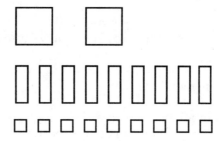

a)   Sketch and label the composite rectangle.

b)   Write the area of the composite rectangle as a sum.

c)   Write the area of the composite rectangle as a product.

d)   Is there more than one way to create a composite rectangle from these tiles? Explain your answer.

AP-30.   Write one polynomial to represent each of the following sums or differences. You may want to draw or visualize algebra tiles to help you understand these problems.

a) $(2x^2 + 3x + 5) + (x^2 + 2x + 8)$

b) $(3x^2 + 8x + 1) + (2x^2 + 8x + 4)$

c) $(3x^2 + 5x + 7) - (4x^2 + x + 1)$

Note: It's okay to get negative results algebraically, even though they can't easily be represented by tiles.

d) $(x^2 + 9x + 8) - (x^2 + 4x + 8)$

e) $(7x^2 + x + 10) - (3x^2 + 12x + 12)$

AP-31.   Solve each of the following equations. Show your steps.

a) $34 = 8 - 2x$ b) $8x - 72 = -8x - 40$

c) $3(x + 2) + 3 = -3$ d) $5(5x + 4) - (x - 3) = 5$

e) $31 = 5 - 2(3x + 4) - x$

f) $\dfrac{2x - 3}{5} = \dfrac{x}{4}$

g) $(x + 5)(x + 2)(x - 1) = 0$ There are three solutions!

AP-32.   Rewrite each of the following expressions by multiplying the binomials in parts (a) and (b) and factoring the polynomials in parts (c) through (f).

a) $(x + 10)(x + 10)$ b) $(x + 2y)(x + 3y)$

c) $9x^2 - 16$ d) $3x^2 - 21x$

e) $x^2 + 3x - 10$ f) $x^2 - 7x - 18$

AP-33.   This graph shows the distances covered by three cars over a certain time interval:

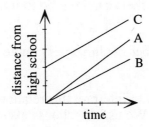

a)   List the cars in order by speed, from greatest speed to least speed, then explain how you know.

b)   Where did cars A and B start? How you know.

AP-34.   A box contains 24 red cubes and B blue cubes. The probability of randomly pulling out one blue cube is $\frac{1}{3}$.

a)   Find B.

b)   Write an equation to represent this problem.

6.3 INVESTIGATING LINEAR RELATIONSHIPS USING GRAPHS

Many of the problems in the remainder of this chapter require several steps in their solutions. Some problems are broken into several tasks and model a process you can use to solve other problems. Be careful to answer <u>all</u> questions and to complete <u>all</u> algebraic expressions, equations, and graphs as specified in each problem. In general, you should solve these problems first from the graph, and then by algebraic methods.

AP-35.   **Cricket Thermometers** Crickets can be used to indicate temperature by using the equation

$$T = \tfrac{1}{4}C + 37,$$

where C represents the number of chirps per minute and T is the temperature in degrees Fahrenheit.

A graph gives us an easy way to interpret this relationship. First you need to decide which variable goes on which axis. We usually choose axes so that the variable on the vertical axis <u>depends</u> on the variable on the horizontal axis. In this case, you are using chirping rate to calculate the current temperature, so the temperature you calculate -- the output -- <u>depends</u> on the chirping rate -- the input.

a)   Before drawing coordinate axes, think about what kind of number it makes sense for C, the number of chirps per minute, to be. Could it be negative? zero? a fraction? Explain why or why not.

b)   Now draw a pair of coordinate axes on graph paper. Scale the horizontal axis so that each tick mark represents five chirps and the vertical axis so that one tick mark represents 10 degrees. Mark each axis up to 100. Be sure to label the axes.

Because it only makes sense for the number of chirps to be a whole number, the Cricket Thermometers graph is *discrete*. It is also true that the Cricket Thermometers equation,

$$T = \tfrac{1}{4}C + 37,$$

is *linear*, so the points on the graph will all lie on a line. It is much easier to graph a line than it is to plot many discrete points, and we can use the **trend line** of the points to help answer questions related to the relationship modeled by the Cricket Thermometers equation.

c)   Use the two-point y-form method, or make a brief table of values, to graph the trend line of the relationship $T = \tfrac{1}{4}C + 37$. If you use just two points, choose them far apart.

d)   Based on your graph, for what temperatures do you think the cricket indicator is not valid? Explain your answer.

e)   What is the approximate temperature when a cricket chirps at a rate of 35 times per minute?

[PROBLEM CONTINUED ON NEXT PAGE]

AP-35. continued

f)   We can represent the statement

"A cricket chirps 35 times per minute."

by drawing a vertical line. Draw the appropriate line so it goes through your graph of the Cricket Thermometers equation.

g)   The equation $T = 50$ means, "The temperature is 50 degrees." Draw a horizontal line to represent this situation. How many chirps per minute would you expect when the temperature is 50 degrees? Draw the appropriate line on the graph and answer the question.

h)   What does the equation $C = 72$ mean? How can it be represented on the graph?

i)   Write another equation similar to $T = 50$ or $C = 72$ and explain in writing what the equation means, just as you did in parts (g) and (h).

j)   Use your graph of the Cricket Thermometers equation to estimate the number of cricket chirps per minute that indicate an increase of one degree in temperature.

AP-36.   **Temperature Conversion** In all countries except the United States, temperature is measured in degrees on the Celsius scale rather than the Fahrenheit scale. The relationship between the two scales can be represented by the equation

$$C = \frac{5}{9}(F - 32),$$

where C gives the degrees in Celsius and F is the degrees Fahrenheit.

On a full sheet of graph paper, draw a pair of axes with the Celsius scale, C, on the vertical axis and the Fahrenheit scale, F, on the horizontal axis. Scale both axes so that one tick mark represents 5°. Make the horizontal axis so that F ranges from –20 to 120 ($-20 \le F \le 120$), and the vertical axis so that C ranges from –30 to 50 ($-30 \le C \le 50$). Finally, graph the relationship $C = \frac{5}{9}(F - 32)$.

a)   Explain in a few sentences how a student from Hong Kong living in the United States could use your graph to convert the temperature from Fahrenheit degrees to the more familiar Celsius scale. Would she first draw a vertical line or a horizontal line? What process could the student use to find the **exact** degrees in Celsius?

b)   Explain in a few sentences how a student from the United States could use your graph to convert temperature readings from Celsius to Fahrenheit. Would he first draw a vertical or a horizontal line? What process could the student use to find the **exact** degrees in Fahrenheit? Describe what would have to be done algebraically, then do it.

AP-37.   To solve some problems -- like the Phone Plan problem in Chapter 3 -- we need to consider two (or more) equations at the same time.

a)   Graph these two equations on the same set of axes:

$$y = -2x + 5$$
$$y = x - 1$$

b)   Locate the point where the two lines intersect and label it "P."

c)   Draw a vertical line through point P to the x–axis.

d)   Draw a horizontal line through point P to the y-axis.

e)   Write the coordinates (x, y) of point P by reading the x value from where your vertical line crosses the x-axis and the y value from where your horizontal line crosses the y-axis.

      Point P is the **point of intersection** of the graphs of the equations $y = -2x + 5$ and $y = x - 1$. It lies on both lines. This means that the x-value and y-value at point P make both equations true; they are a common solution to the two equations. Verify this, and then update your Tool Kit.

We can sometimes find the point that makes two equations true simultaneously (at the same time) by carefully graphing each equation. Most of the time, however, it is very difficult to find an exact answer by graphing; the graphs give only close approximations of the x- and y-values we seek. Algebraic techniques will produce exact results, though.

AP-38.   In AP-37, you found the point of intersection of the graphs of $y = -2x + 5$ and $y = x - 1$ by graphing the two equations. You can also use algebraic methods to find the one pair of numbers (x, y) that makes both equations true. The problem is to find a number x that will give the same result for y in each of the equations:

$$y = -2x + 5$$
$$y = x - 1$$

You could use guess and check, but there is a faster, easier way. Notice that if the results for y are the same number, then the expressions $-2x + 5$ and $x - 1$ must both represent that number since they are both equal to y. So you can write:

$$-2x + 5 = x - 1.$$

You've solved equations like this before, so you've taken a new problem and turned it into something familiar.

a)   Solve the equation $-2x + 5 = x - 1$ for x.

b)   In part (a) you found the x-coordinate of the point of intersection. You also need to find the y-coordinate. To do this, substitute the solution for x into $y = -2x + 5$.

c)   Repeat part (b), this time substituting the solution for x into $y = x - 1$. What do you notice about the results?

d)   Write the coordinates of the point where the graphs of the two equations intersect.

e)   Compare the two methods -- graphing and algebraic -- for finding the point of intersection of two lines. Which seems easier for you? Why?

AP-39.   Use the algebraic procedure in problem AP-38 to find the coordinates of the point where each
         of these pairs of lines (in y-form) intersect:

a) $y = -x + 8$
     $y = x - 2$

b) $y = -3x$
     $y = -4x + 2$

c) $y = -x + 3$
     $y = x + 3$

d) $y = -x + 5$
     $y = \frac{1}{2}x + 2$

e)   Graph the equations in part (a) to check your answer.

AP-40.   Transform (change) each of the following equations into y-form and then graph them using the
         two-point y-form method.

a) $y - 4x = -3$

b) $3x + 2y = 12$

c) $3y - 3x = 7$

AP-41.   Solve each of the following equations. Show your steps.

a) $5 + 4(x + 1) = 5$

b) $\frac{2x}{3} - 5 = -13$

c) $4(2x - 1) = -10(x - 5)$

d) $4(x + 5) - 3(x + 2) = 14$

e) $\frac{2x}{7} = \frac{4}{5}$

AP-42.   Without drawing a graph, determine which, if any, of the points are on the graph of the given
         linear equation.

a) $(8, 4)$, $(4, 0)$, $(4, 3)$, $(3, 4)$ for $y = \frac{1}{4}x + 2$

b) $(0, -7)$, $(2, -1)$, $(4, 5)$, $(6, 11)$ for $y = 3x - 7$

AP-43.   Find x and y for each pair of similar triangles below.

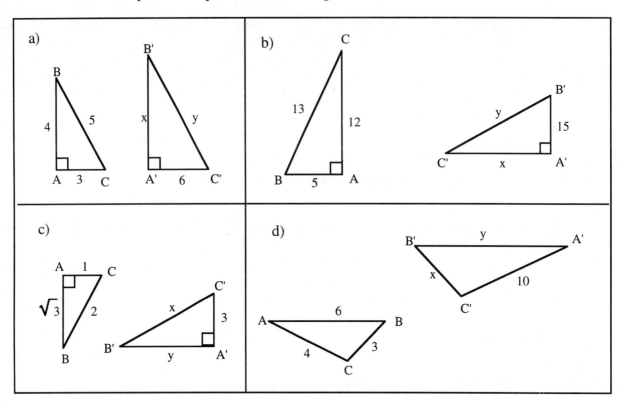

AP-44.   Rewrite each of the following expressions by multiplying the binomials in parts (a) and (b) and
          factoring the polynomials in parts (c) through (f).

a) $(2x + 3)(x + 5)$ b) $(4 - x)(3x + 10)$

c) $x^2 - x - 6$ d) $x^2 - 121$

e) $x^2 + 8x - 20$ f) $6x^3 - 30x^2 + 12x$ Look for common factors.

6.4 SOLVING SYSTEMS OF LINEAR EQUATIONS BY THE SUBSTITUTION METHOD

AP-45.   **The Amusement Park** Jim and his friends are going to the amusement park and find that they have two ticket options. In one option each person could buy an admission ticket for $25.50 and then pay 85¢ for each ride. The other option is to buy an admission ticket for $11.90 and then pay $2.55 a ride. What do you think Jim should do?

a)   How much does Jim spend if he buys the $25.50 admission ticket and goes on four rides?   12 rides?   15 rides?

b)   How much does he spend if he buys the $11.90 admission ticket and goes on four rides?   12 rides?   15 rides?

c)   Let x represent the number of rides and y represent the total amount spent (in dollars). Write an equation for <u>each</u> ticket option. Use the patterns from parts (a) and (b) to help you.

d)   Explain why the graph of each ticket option equation is discrete. Then use parts (a) and (b) or the equations from part (c) to graph the trend line for each ticket option. Scale the x-axis one ride per unit, for up to 15 rides. Scale the vertical axis two dollars per unit, for up to $52.00.

e)   Use the graphs from part (d) to estimate the number of rides that give the same total cost regardless of the option Jim chooses. Estimate the total cost that corresponds to this number of rides.

f)   Use your equations from part (c) to solve for the number of rides that give the same total cost, and then find the associated cost.   Refer to problem AP-38 if you need help. Compare your answers to your estimates in part (e).

g)   Now conclude the problem by writing two or three sentences to advise Jim as to what he should do. (Several responses are possible. You just need to give one.)

AP-46.   **Going Shopping** Barb and Betsy went shopping. Barb went to Aardvark Records and bought some tapes. Afterwards she spent $12 on a sweatshirt. Betsy bought some tapes and she also bought a pair of earrings for $3. We don't know the number of tapes each one bought, the price of each tape, or the total amount each one spent. Let's assume the tapes all cost the same amount.

a)   Let x represent the price of a tape (in dollars) and let y represent the total amount of money each person spent (also in dollars).

1)   Suppose Barb bought three tapes. Write an equation relating x and y to show the <u>total</u> <u>amount</u> of money she spent.

2)   Suppose Betsy bought five tapes. Write another equation relating x and y to show the <u>total</u> <u>amount</u> of money Betsy spent.

b)   Are the graphs of the expenditure equations in part (a) discrete? Explain why or why not.

[PROBLEM CONTINUED ON NEXT PAGE]

AP-46. continued

c) Draw a set of coordinate axes the x-axis scaled by two units per dollar, for up to $8.00, and the y-axis scaled by one unit per two dollars, for up to $30.00. Graph the trend line for each of the equations from part (a) and label each with its equation. Extend the lines beyond their point of intersection.

d) Find the coordinates of the point of intersection of the two lines. What is the real-world meaning of the point of intersection? Reply in a complete sentence.

e) Suppose Barb and Betsy each spent the same total amount of money. This means that both y-values are the same. Now you can figure out how much a single tape costs. Do so in two ways:

(1) Estimate the cost from the graph, and

(2) use your equations from part (a) to write one equation involving x, and then solve for x.

AP-47. **The Substitution Method for Solving Pairs of Linear Equations**
In problems AP-37 and AP-38 we found a single solution for two equations in y-form by making one equation from the two and eliminating a variable. The same thing can be done even if only one equation is in y-form (or "x-form").

This example shows the **substitution method** for finding a common solution to a pair of linear equations:

$$y = x - 1$$
$$x + y = 11$$

Just as in AP-38, there are two equations and two variables:

$$y = x - 1$$
$$x + y = 11$$

Once again we can substitute for (replace) y with something that is equal to it. The first equation tells us that y is equal to x – 1, so we can replace the y in the second equation with x – 1:

$$x + y = 11$$
$$x + (x-1) = 11$$

The resulting equation looks like many we have solved before:

$$2x - 1 = 11$$
$$2x = 12$$
$$x = 6$$

Remember that we are finding the point where the two lines cross, so we must find the "y" value also. We can use either equation to obtain y = 5:

$$y = x - 1 = 6 - 1 = 5$$
$$\text{or } 6 + y = 11, \text{ so } y = 5$$

a) If you had graphed the lines y = x – 1 and x + y = 11, where would they cross?

b) Use the **substitution method** to solve this pair of equations:
$$y = x + 3$$
$$x + 3y = 5$$
(If you need help, try substituting into the second equation, x + 3(_____) = 5.)

AP-48.   Use the substitution method to find the point of intersection (x, y) for each pair of linear equations below.

    a) $y = -3x$
         $4x + y = 2$

    b) $2x + 3y = -17$
         $y = x - 4$

    c)   Show both of the checks that your solution to part (a) is correct.

AP-49.   Use the substitution method to find the point of intersection (x, y) for each pair of linear equations below.

    a) $x + y = 4$
         $x = y - 2$

    b) $y = x - 3$
         $3x - y = 8$

    c) $y = 2x - 3$
         $x - y = -4$ Hint: $y = (2x - 3)$.

    d)   Show both of the checks that your solution to part (a) is correct.

AP-50.   Solve each of the following equations for x, if possible.

    a) $4 + 2.3x = -5.2$

    b) $2(4x - 7) = 8x + 14$

    c) $6(x - 2) = 5(x - 11) - 21$

    d) $3(x - 5) = \frac{1}{5}(10x - 25)$

    e) $\dfrac{x}{x + 4} = \dfrac{9}{2}$

    f) $x(2x - 3) = 0$

AP-51.   Rewrite each of the following expressions by multiplying the factors in parts (a) and (b), and by factoring the polynomials in parts (c) through (e).

    a) $(2x^2 + 4)(5x - 9)$

    b) $(x + 2)(2x^2 - 3x + 6)$

    c) $x^2 - 0x - 64$

    d) $x^2 - 6x - 72$

    e) $24 - 16x$

AP-52.   Write an equation for the following problem and then solve the equation.

Mai started on the bike trail at 10 miles per hour. Ly left one hour later at 15 miles per hour.
How long does it take before Ly catches up with Mai?

AP-53.   Complete each of the following diagrams and write an equation to express "the area as a
product equals the area as a sum."

a)

x^2	$3x$
$2x$	6

b)

$4y^2$	
	-9

AP-54.   a)   Graph the line $y + x = 5$.

b)   Slide the entire line one unit to the right and down three units. Draw the new line.

c)   Describe your graphs.

6.5 SYSTEMS OF NON-LINEAR EQUATIONS

AP-55.   Make a table and graph each of the following lines on the same coordinate system. Label each line with its equation.

   a) $y = 4$ b) $x = -2$

   c) $y = 0$ d) $x = 0$

AP-56.   *Without graphing,* write a sentence to describe each of the following lines so that someone who is absent today could graph the lines from your descriptions.

   a) $y = 6$

   b) $x = 3$

AP-57.   a)   Marci correctly graphed an equation and got a vertical line that contained the point $(1, 5)$. What equation did Marci graph?

   b)   When Marc graphed his equation he got a horizontal line that contained the point $(1, 5)$. Assuming that the graph was correct, what equation did Marc use?

AP-58.   ➔**Estimating Rocket Height** A rocket is launched and its distance above sea level is recorded until it lands. A graph of the data is shown below. Use your copy of the AP-58 resource page to complete the following examination of the graph.

   a)   Estimate the rocket's height above sea level 3.5 seconds after firing. Draw a vertical line on your copy of the graph and show its point of intersection with the curve.

   b)   Estimate how long after being launched the rocket is 250 feet above sea level. Draw a horizontal line on your copy of the graph and show its points of intersection with the curve.

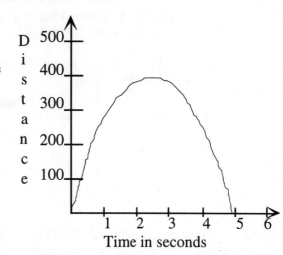

   c)   Write the equation of the vertical line in part (a) and the equation of the horizontal line in part (b).

AP-59.   You have found solutions to systems of linear equations in two ways: by graphing both
         equations and finding the point of intersection of the graphs; and by using the method of
         substitution to solve the equations algebraically. You can also find solutions to systems that
         contain equations which are not linear.

         Graph each pair of equations below on the same coordinate axes. In this problem, at least one
         of the two equations does not produce a straight line. You will have to use your calculator to
         make a table of values to help you graph these curves. Be careful with signs, especially $-x^2$.
         Estimate the point(s) of intersection within 0.5 unit.

         It may surprise you that it is impossible to solve some of these problems, such as part (a), by
         algebraic methods. Devising computer programs to find approximate solutions to such
         problems has become a very important mathematical activity.

         a) $y = 2^x$ for $x = -2, -1.5, -1, \ldots , 3.5, 4$
              $y = 3x + 1$ for $-2 \le x \le 4$

         b) $y = x^2 - 4x$
              $y = -x^2 + 2x + 1$, both using $-3 \le x \le 5$

AP-60.   a)   On a new set of axes, graph both $y = x^2$ and $y = 4$.

         b)   Give the coordinates where the line crosses the parabola.

AP-61.   Kelley and Jamilla were both trying to solve the equation $x^2 = 9$. Jamilla used the square root
         key $\boxed{\sqrt{}}$ on her calculator and got $x = 3$, but then she noticed that Kelley had $x = -3$ written
         on her paper. After checking that both $x = 3$ and $x = -3$ were correct, they remembered that
         some equations have two solutions.

         Each of the following equations also has **two solutions**. Find both solutions and round each
         one to the nearest 0.01.

         a) $x^2 = 4$ b) $x^2 = 25$

         c) $s^2 = 30$ d) $z^2 = 148$

         e) $y^2 - 1 = 15$ f) $t^2 + 4 = 24$

AP-62.   Compare problems AP-60(b) and AP-61(a). Why do you think that the x-values are the same?

AP-63.   Sonja notices that the lengths of the three sides on a right triangle are consecutive even
         numbers. The triangle's perimeter is 24. Write an equation and solve it to find the length of
         the longest side.

AP-64.   Solve each system of equations for x and y (that is, where do the lines intersect?).

   a) $y = x - 1$ b) $y = 3x - 8$
        $x + 2y = 4$ $y = 2x + 7$

   c) $x = 2y$ d) $y = 3$
        $x + 3y = 10$ $2x + y = -7$

   e) $x = -2$ f) $y = 3x - 8$
        $2x + y = -7$ $2x + 3y = 12$

AP-65.   You remember that $-x^2 \neq (-x)^2$. How can you convince Vince, a new student, that they are
          different? Show the difference with numbers, diagrams, or some other approach.

AP-66.   Write each of the following exponential expressions in a simpler form.

   a) $(2x^5)^2$

   b) $\dfrac{3x^3y^2}{6xy^5}$

   c) $(5.32 \cdot 10^{15})^2$

   d) $\sqrt{\dfrac{2.1 \cdot 10^{15}}{4.2 \cdot 10^6}}$

AP-67.   Factor each of the following quadratics. You may want to draw generic rectangles.

   a) $x^2 + 4x + 3$ b) $x^2 - 7x + 12$

   c) $x^2 + 10x + 16$ d) $x^2 + 5x - 24$

   e) $x^2 - 3x - 18$ f) $x^2 - 121$

AP-68.   a)   Graph the line $y = 2x - 5$.

          b)   Slide the line up two units, then draw the new line. If you have trouble, slide two of the
               points on $y = 2x - 5$ up two units each, then draw the new line.

          c)   What is the solution for the system of equations shown by these two lines?

6.6 SOLVING QUADRATIC EQUATIONS: GRAPHING AND THE ZERO PRODUCT PROPERTY

You've seen that you can sometimes use graphing to solve a system of linear equations, although algebraic methods are more accurate, and often easier. In this section you'll examine the solution of quadratic equations. First you'll write a quadratic equation by using substitution in a system of equations where one of the equations is quadratic (of the form $y = ax^2 + bx + c$) and the other equation is $y = 0$. To solve such a system, you'll initially graph both equations and find the points of intersection. Then you'll see how you an use factoring and the zero product property to solve quadratic equations algebraically. This approach provides a way to solve the original system.

AP-69. a)   Graph the parabola $y = x^2 + 4x - 5$ for $-6 \le x \le 5$. On the same set of axes, graph the line $y = 0$. Make sure that everyone in your group has the same graph.

b)   Where does the line $y = 0$ intersect the parabola $y = x^2 + 4x - 5$?

c)   Use substitution to write a single equation from the two equations in part (b). (Look back at AP-47 if you need help.)

d)   What are the solutions to the equation you wrote in part (c) ? Describe how you got your answers.

AP-70.   By now you have factored quite a few quadratic expressions similar to $x^2 + 2x - 3$. You have also graphed several parabolas in "y-form" where a quadratic expression appears on the right side of the equation, such as $y = x^2 + 2x - 3$, for example. Now you'll to put these two ideas together by using the **zero product property**.

a)   Factor $x^2 + 2x - 3$.

b)   Graph $y = x^2 + 2x - 3$ for $-4 \le x \le 3$.

c)   Estimate the coordinates of the points where the graph in part (b) crosses the x-axis. (These points are called the **x-intercepts** of the graph.)

d)   What are the values of y at the points where the graph crosses the x-axis? What is the equation of the x-axis? (Hint: The x-axis is a horizontal line.)

e)   Describe how the zero product property helps you solve the equation $(x + 3)(x - 1) = 0$. (Refer to your Tool Kit (or re-read TR-46) if you need help.) Then use the zero product property to solve $(x + 3)(x - 1) = 0$ for x.

f)   Compare your results in part (e) to the x-values you estimated in part (c). What do you notice? Now write a sentence or two to describe the relationship between the results of the two methods you used:
   finding where the <u>graph</u> of the equation $y = x^2 + 2x - 3$ crosses the x-axis, and
   solving the equation $0 = x^2 + 2x - 3$ using algebra.

If you haven't already done so, add "x-intercept" and "zero product property" to your Tool Kit.

AP-71. a) Graph the parabola $y = x^2 - x - 2$ for $-2 \leq x \leq 3$. Mark the x-intercepts (points where the graph crosses the x-axis) and label them with their coordinates.

   b) Substitute $y = 0$ into the equation $y = x^2 - x - 2$ to obtain the single equation. Then find two solutions to the new equation by first factoring the expression $x^2 - x - 2$.

   c) Explain how solving the equation $0 = x^2 - x - 2$ gives you enough information to name the x-intercepts of the parabola $y = x^2 - x - 2$ without having to draw the graph.

   d) Why might some people call the method you followed in parts (b) and (c) the "zero product" technique?

AP-72. Without graphing, use the zero product technique developed in the previous problems to find where the each of the following graphs crosses the x-axis. (Hint: What is the equation of the x-axis?)

   a) $y = (x - 3)(x - 4)$

   b) $y = x^2 + 5x - 24$

   c) $y = x^2 + 10x + 16$

AP-73. Solve each of the following quadratic equations.

   a) $0 = (x + 3)(x - 6)$

   b) $x^2 + 4x - 32 = 0$

   c) $0 = 2x^2 + 7x + 6$

   d) Substitute the two values that you found for x in part (a) into the equation to show that both values are correct solutions.

   e) Find the coordinates of the x-intercepts of the parabola $y = x^2 + 4x - 32$ without graphing. (Hint: Use what you found in part (b).)

AP-74.   (a)   Jamilla and Kelley wondered if they could use the zero product method to solve the equation $x^2 = 64$, so they first wrote $x^2 - 64 = 0$, and then tried to factor $x^2 - 64$. Here's how their paper looked:

$$x^2 = 64$$
$$x^2 - 64 = 0 \qquad \underline{\hspace{5cm}}$$
$$x^2 + 0x - 64 = 0 \qquad \underline{\hspace{5cm}}$$
$$(x + 8)(x - 8) = 0 \qquad \underline{\hspace{5cm}}$$
$$x + 8 = 0 \quad \text{or} \quad x - 8 = 0 \qquad \underline{\textit{Used the zero-product property}}$$
$$\text{So, } x = -8 \quad \text{or} \quad x = 8 \qquad \underline{\hspace{5cm}}$$

Copy Jamilla's and Kelley's work and describe what they did in each step. Then check that both solutions are correct.

Now use the zero product method to solve each of the following quadratic equations. Check your solutions.

b) $x^2 = 169$ c) $z^2 - 0.7 = 0.57$

d) $y^2 - 0.1 = 1.11$ e) $t^2 + 100 = 2600$

AP-75.   Solve this system of non-linear equations by drawing careful graphs, as you did in AP-59:

$$y = \frac{3}{x}$$
$$y = 3^x, \quad \text{both using } x = -2, -1.5, -1, \dots, 2, 2.5$$

AP-76.   Plot the points A $(0, 0)$, B $(3, 0)$, and C $(3, 2)$. Connect the points to form $\triangle ABC$. Slide $\triangle ABC$ three units to the right and up two units. Then draw the new triangle and describe the results of the slide. What are the coordinates of the new vertices?

AP-77.   When the employees of the accounting firm of Scrooge & Marley were asked how long they'd been with the firm, the responses were recorded on a graph. Each x on the graph below represents one employee.

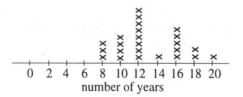

a)   Calculate the average number of years an employee has worked for Scrooge & Marley

b)   What percent of the employees has been with Scrooge & Marley at least twelve years?

c)   If you had to choose one of the employees at random to interview, what is the probability that the chosen person worked less than twelve years for Scrooge & Marley?

AP-78.   Factor each of the following quadratic polynomials. (It may help to draw generic rectangles.)

a) $2x^2 - 6x$

b) $x^2 - 4x + 3$

c) $2x^2 - 11x + 5$

AP-79.   Solve each equation by any method you choose.

a) $2x + 6 = 3x - 4$

b) $9s - 1 = 4$

c) $32 = \frac{1}{7}r + r$

d) $(x - 7)(2x + 6) = 0$

e) $3(x + 7) = 3x + 21$

f) $x^2 - 4x - 12 = 0$

g) $2x(x - 5) = 0$

h) $1.5(w + 2) = 3 + 2w$

AP-80.   Solve this system of linear equations:

$$y = 3x - 1$$
$$x + 2y = 5$$

AP-81.   Find the dimensions of each generic rectangle. Write an equation that expresses the area of each generic rectangle as a product of its factors.

a)

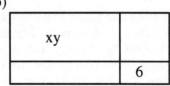

Compare your solutions for parts (a) and (b) with those of other members of your group. Write one or two sentences to describe what you notice.

AP-82.   **Chapter 6 Summary: Rough Draft** The Introduction to the chapter mentioned five main ideas that would be developed in the problems you did. Re-read the Introduction and then look back through the chapter to find where each main idea was developed

Write your answers to the following questions in rough draft form **on separate sheets of paper**, and be ready to discuss them with your group at the next class meeting. Focus on the **content**, not neatness or appearance, as you write your summary draft. You will have the chance to revise your work after discussing the rough draft with your group.

a)   In this chapter you've worked on many problems that relate graphs and equations. From your work, select two problems that best show your understanding of how graphs and equations are related. Write complete sentences to describe how you did each of the selected problems. Then tell why you chose the problems that you did.

b)   Select a problem you didn't understand before, but now know how to do. Show all your work and a complete solution. Explain why you chose the problem.

6.7 SUMMARY AND REVIEW

AP-83.   **Chapter 6 Summary: Group Discussion**   Take out the rough draft summary you completed in AP-82. Use this time to discuss your work and use homework time to revise your summary as needed.

Take turns to describe the problems group members chose to illustrate their understanding of how graphs and equations are related. Each group member should:

- explain the problems he or she chose to illustrate the main ideas;

- explain why he or she chose those particular problems; and

- explain a problem he or she didn't understand before, but now can solve, and explain why the problem was chosen.

This is your chance to make sure your summary is complete, update your Tool Kits, and work together on problems you may not be able to solve yet.

AP-84.   Carol and Christina leave from the same place and travel on the same road. Carol walks at a rate of two miles per hour. Carol left five hours earlier than Christina, but Christina bikes at a rate of six miles per hour. Suppose we want to use algebra to find out how long it will take for Christina to catch up with Carol.

a)   We are told that Christina catches up with Carol, so we know they have traveled the same distance by that time. The diagram shows this with equal-length arrows. Start your solution by writing, "Let x represent the time (in hours) that Christina travels" on your paper. This will be the amount of time it takes Christina to catch up with Carol. Then copy the diagram on your paper.

Carol ———————————————————▶

Christina ————————————————▶

b)   Write the length of time and the rate Carol is traveling on your diagram. Do the same for Christina. Be sure to write units each time you write a number (don't just write "5"; write "5 hours").

c)   Write an expression using x to represent the time Christina travels. On your paper, write "Carol's distance = _____ " so that you know what the expression represents.

d)   Write an expression using x to represent the time Christina travels. Be sure to write "Christina's distance = _____ ."

e)   Write an equation that says Carol and Christina travel the same distance, and then solve the equation.

f)   How long does it take Christina to catch up with Carol?

AP-85.   **Ant Thermometers** A certain kind of ant can be used as a thermometer! The ant travels faster as the temperature increases. To calculate the temperature we can use the equation

$$T = 11S + 39,$$

where S represents the ant's speed in inches per second and T represents the temperature in degrees Fahrenheit.

a)   Graph this rule with the temperature, T, on the vertical axis and the ant's speed, S, on the horizontal axis. Scale the vertical axis by five degrees per tick mark, up to 100 degrees. Scale the horizontal axis so that four units of graph paper represent a speed of one inch per second, and mark it to six inches per second.

b)   For what temperatures is the temperature equation impractical?

c)   From your graph, estimate the temperature when the ant is traveling at a speed of two inches per second by drawing a vertical line at $S = 2$. Record your estimate.

d)   Draw a line on your graph to represent a temperature of 90 degrees. What is the ant's speed when the temperature is 90 degrees?

e)   Use your graph from this problem and from the Cricket Thermometers problem, AP-35, to answer the following question:

      If a cricket chirps 108 times per minute, what is the approximate speed of the ant?

AP-86.   **More or Less** Judy has $20 and is saving at the rate of $6 per week. Jeanne has $150 and is spending at the rate of $4 per week. After how many weeks will each have the same amount of money?

Solve this problem in a step by step manner by writing an equation for each person and drawing the graphs to estimate the solution. (Scale the x-axis by one week per tick mark, up to 20 weeks. Scale the y-axis by $10 per tick mark, up to $160.) Then show an algebraic method for finding the solution exactly.

AP-87.   Graph the equation $y = -3x + 4$ using the two-point y-form method.

AP-88.   Without drawing a graph, determine which of the points listed below are on the graph of the linear equation $y = -2x + 3$.
                    $(0, 3)$, $(3, 0)$, $(-3, 3)$, $(1, 1)$, $(5, -7)$, and $(5, 7)$.

AP-89.   Solve each of the following equations for x. Show all your work.

a) $2x - 3 = 0$ b) $0 = (2x - 3)(x + 5)$

c) $0 = 2x^2 + 7x - 15$ d) $2x^2 + 7x - 15 = -21$

e) $8(x + 6) + 23 = 7$ f) $\dfrac{5 - 2x}{3} = \dfrac{x}{5}$

g) $3x - 11 = 0$ h) $Dx - C = 0$

i) $x^2 - 4x + 4 = 0$ j) $x^2 + 6 = 6$

AP-90.   Solve the each of the following systems of equations by any method(s) you choose.

a) $y = 3x + 1$
      $x + 2y = -5$

b) $x = y + 3$
      $x + 2y = -6$

c) $y = x^2 - 5x + 4$
      $y = 0$

l) $y = x^2 - 3x$
      $y = 2x - 4$

AP-91.   Find the x-intercepts for the graph of each of the following equations.

a) $y = x^2 + 10x + 21$

b) $y = 2x - 1$

c) $y = 2x^2 - 5x + 2$

AP-92.   The area of this generic rectangle is 60 square units:

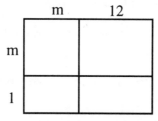

a)   Write an equation, clean it up, and then solve to find the value(s) of m.

b)   Can there be more than one value for m? Explain why or why not.

AP-93.   **Chapter 6 Summary: Revision** Use the ideas your group discussed in AP-83 to revise the your rough draft of a Chapter 6 Summary. Your presentation should be thorough and organized, and should be done on paper separate from your other work.

AP-94.   **Course Summary** It may seem strange that this problem is called "Course Summary" when you haven't even completed the course yet. But the idea of writing a summary of what you've learned so far in the course is really not so strange. Recall that the idea of writing a chapter summary was introduced in Chapter 1 by this statement:

> *In this course, as in most courses you study, you will continue to work on many ideas, concepts and skills. There is a lot of information in this course. It may be very difficult to process this information and organize it for yourself, so that you can easily remember **all** of it. To help remember information, it helps to summarize it in the context of big ideas. You will be asked to write a summary for every chapter in this course. This is an important process in learning.*

A summary of the main ideas you've learned in the course so far will help you see how much you've learned and how it all fits together. Use your chapter summaries to identify the important ideas you've learned. List at least five, and no more than eight, important ideas and choose one or two problems which represent each idea.

6.8 THE LUNCH BUNCH INVESTIGATION (OPTIONAL)

AP-95.   **The Lunch Bunch (A Mathematical Play)** Read this play with your classmates to uncover an interesting mathematical problem with a surprising solution.

One stormy day in the middle of winter, a bunch of Ms. Speedi's students got together to discuss everything under the sun (or rather, that day, under the clouds). After they had spent about half an hour talking, Smarty Marty asked the group a question.

> Smarty Marty: "Can anybody think of two numbers whose product equals their sum?"

About ten seconds elapse. Speedy Stevie thinks he knows the answer.

> Speedy Stevie: "Ah, that's too easy. They're just zero and zero."

> Smarty Marty: "Hey, that <u>was</u> too easy. Zero is special in multiplication. Find two <u>positive</u> numbers that work."

All of a sudden, Logical Lucy has a huge smile on her face.

> Logical Lucy: "I know another pair that works, and I'm not telling no one."

> Patty Pattern: "Why won't you tell us, Lucy?"

> Logical Lucy: "It took a lot of time to think of it, and if I tell you, you won't appreciate all of thetime I put into it."

> Dubious Dan: "I bet you she doesn't know."

> Logical Lucy: "I do, too. It's two and two. (*Pause.*) I can't believe it! I fell for the oldest trick in the book!"

> Smarty Marty: "That has to be all of the answers. A problem can't have more than two answers, right guys?"

> Dubious Dan: "Hey, you know Stevie is right; zero and zero do work."

> Careful Carrie: "Wait a second, guys. I think I know another pair that works, and it's strange. Check this out: three and one point five (1.5) works."

[PROBLEM CONTINUED ON NEXT PAGE]

AP-95. continued

Smarty Marty:	Yeah, you're right, 3 times 1.5 equals 4.5, just like 3 plus 1.5 equals 4.5."
Patty Pattern:	"You guys, I don't see a pattern. Why do 0 and 0 work, 2 and 2 work, and 3 and 1.5 work? It doesn't make any sense. Carrie, can you think of another pair?"
Logical Lucy:	"I know, but I'm not telling."
Dubious Dan:	"You don't know nothing. You're probably lying again."
Logical Lucy:	"Again!!! What do you mean?! Try five and one point two (1.2). (*Looks angry.*) I can't believe I fell for it again!"
Smarty Marty:	"That doesn't work, Lucy, ... but 5 and 1.25 works."
Careful Carrie:	"Hey, that's pretty good."
Patty Pattern:	"Now I think I see a pattern!"

Can you find the pattern that Patty sees? In this investigation you'll use your graph interpretation and equation-solving skills to answer the question Marty posed.

AP-96.   **Another Solution?** The Lunch Bunch wants to find out whether there are any pairs of positive numbers whose sum is six and whose product is six. Dubious Dan says, "Let's try lots of guesses." But Logical Lucy says, "How about graphing? It's easier."

a)   Graph the line $x + y = 6$.

b)   On the same coordinate axes, graph $xy = 6$.

c)   Approximately where do the two graphs meet?

d)   Explain in one or two sentences how graphing can be used to find two numbers that have a particular sum and product.

AP-97.   **The Investigation Continues** Carrie wonders if there are two positive numbers whose sum is her favorite number and whose product is also her favorite number.

a)   Choose some variables and use them to write two equations to represent Carrie's problem.

b)   Use the two equations you found in part (a) to write a single equation.

c)   Look back to the Lunch Bunch play at the beginning of this section. Compare Marty's opening question to your equation in part (b). Describe what you notice.

AP-98.   **The Lunch Bunch Returns!** If you were to graph the equation you wrote in part (b) of the previous problem, do you think it would be a straight line? Here's what the Lunch Bunch thinks:

Speedy Stevie: "I think it's a line."

Smarty Marty: "I'm not so sure. Let's make a table and fill in what we know."

Patty Pattern: "Maybe, if we change all the decimals to fractions we will see a pattern."

Logical Lucy: "Hey, I see some symmetry here."

Dubious Dan: "There she goes again. She doesn't know what she's talking about!"

Logical Lucy: "I do so. Look, we can go back and forth from x to y. We know the pair 5 and 1.25 works, right? Well, if we let x equal 5, then y equals 1.25. And if we let y = 5, then x = 1.25. See?! We can get two table entries by knowing one pair of numbers that works."

Careful Carrie: "Those are all great ideas and here's another one. We can use the equation to find more number pairs. Just choose any value for x and replace x in the equation with that value.

Look, if I choose 4 for the value of x, then
$$xy = x + y \text{ becomes}$$
$$4y = 4 + y.$$
Now we just need to solve for y. Let's see ... *(Pause.)*

Speedy Stevie: "I got it! We have $4y = 4 + y$, so $3y = 4$. That means $y = \frac{4}{3}$ or about 1.33 when x is 4 !"

a)   Copy and complete the following tables. To fill in the tables, use all of the pairs of numbers the Lunch Bunch has already found and all of their ideas from the discussion you read above. Also use some ideas of your own!

x	1	1.1	1.125	1.16	1.25	1.3	1.5	2	3	4	5
y											
Fractional form of y											

x	6	7	9	10	11
y					
Fractional form of y					

b)   Now use the tables to draw a graph. What happened when you replaced x with the value 1 ? What would y equal if x were replaced by 1.01 ?

AP-99.   Do you think it is possible to find any negative numbers that would work in the Lunch Bunch's problem? Is there more to the graph in AP-98 than the Lunch Bunch thought? If possible, make a table of values for x < 1 and use the values to draw the rest of the graph.

AP-100.   Write a short paragraph to summarize what you learned in the Lunch Bunch investigation about word problems, graphs, and systems of equations.

AP-5. The Bike Race

Cindy and Dean are engaged in a friendly 60-mile bike race that started at noon. The graph below represents their race. Notice that Dean, who doesn't feel a great need to hurry, stopped to rest for a while.

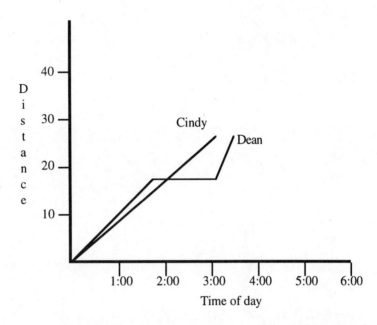

To answer each of the following questions, show how to use the graph, and then write your responses.

a)   At what time (approximately) did Cindy pass Dean?

b)   About how far has Cindy traveled when she passes Dean?

c)   You can see that Dean will catch up to Cindy again if they both continue at their same rates. At approximately what time will this happen? About how far have they both traveled to this point?

d)   Does Dean travel faster or slower after his rest?

e)   About how long did Dean rest?

f)   If Cindy continues at a steady pace, about how long will it take her to complete the race?

Algebra Tool Kit: the "what to do when you don't remember what to do" kit

Algebra Tool Kit: the "what to do when you don't remember what to do" kit

AP-58. ➤Estimating Rocket Height

A rocket is launched and its distance above sea level is recorded until it lands. A graph of the data is shown below.

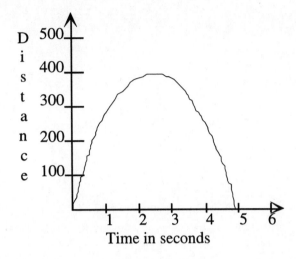

a)   Estimate the rocket's height above sea level 3.5 seconds after firing. Draw a vertical line on your copy of the graph and show its point of intersection with the curve.

b)   Estimate how long after being launched the rocket is 250 feet above sea level. Draw a horizontal line on your copy of the graph and show its points of intersection with the curve.

c)   Write the equation of the vertical line in part (a) and the equation of the horizontal line in part (b).

Chapter 7

THE BUCKLED RAILROAD TRACK:
Words to Diagrams to Equations

CHAPTER 7

THE BUCKLED RAILROAD TRACK:
WORDS TO DIAGRAMS TO EQUATIONS

Mathematical problems that arise outside of a mathematics class rarely come in the form of an equation that's ready to be solved. Usually mathematical problems arise in a context and are stated in words, so they more closely resemble "word problems" than straight forward "solve for x" type problems. When trying to solve a word problem, it is often helpful to draw a picture or a diagram to represent the information in the problem. Many times the diagram can then be used to help write an equation that models the problem. Solving the original problem then becomes a matter of solving an equation and then checking that the solution (or solutions) makes (make) sense in the problem. In this chapter you'll be increasing your problem solving and algebra skills as you work on problems similar to this one about a buckled railroad track:

A new railroad line was installed in the Central Valley. In order to reduce derailments along a two-mile stretch, the track was made with straight one-mile long rails. The rails were laid in the winter and they expanded in the heat of the following summer. Indeed, each mile-long rail expanded one foot in length! Ordinarily, because the rails do not bend, they would jut to the side. However, in this strange case, the rails jutted upward where their ends met. How high above the ground were the expanded rails at the joint?

In this chapter, you will have the opportunity to:

- develop your skill at drawing diagrams to make it easier to write equations that can be used to solve problems posed originally in words;

- use the Pythagorean relationship in right triangles to write quadratic equations;

- develop understanding of some properties of square roots of numbers; and

- use the idea of a "fraction buster" to convert an equation with fractions into an equivalent equation without fractions.

CHAPTER CONTENTS

7.1 RIGHT TRIANGLES AND THE PYTHAGOREAN RELATIONSHIP

RT-1.   When Casey heard the Buckled Railroad Track problem he tried to draw a sketch of the situation using smaller numbers. He reasoned that if the rail were originally 4 centimeters long and it expanded one centimeter to 5 centimeters long, then the jutting end would be 1 centimeter above the ground.

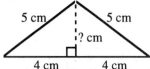

Make an accurate drawing to check Casey's conjecture. Measure the length of the third side of each right triangle to the nearest 0.1 centimeter and record the measurement. Is the height of each triangle one centimeter? What does this tell you about the length of the third side of a right triangle when the longest side is one centimeter longer than the base?

RT-2.   In a right triangle, the longest side is called the **hypotenuse**. The two shorter sides, which form the right angle, are called **legs**. Add this information, along with a labeled diagram, to your Tool Kit.

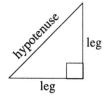

RT-3.   **Sides of Right Triangles** Apparently the relationship among the sides of a right triangle is not as simple as Casey first thought. In the previous chapter you worked with quadratic and other, more complicated relations.

a)   How complicated is the relationship among the legs and hypotenuse of a right triangle? To help answer this question, on the resource page for RT-3 complete the table for the five right triangles.

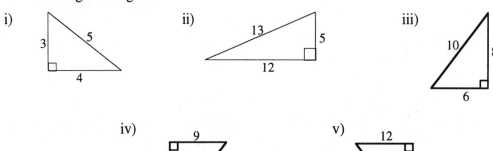

	length of leg#1	length of leg#2	length of hypotenuse	(length of leg#1)^2	(length of leg#2)^2	(length of hypotenuse)^2
i)						
ii)						
iii)						
iv)						
v)						

RT-3. continued

    b)   Look for a pattern in the three columns on the far right of your table in part (a). Describe what do you observe by completing this statement:

"In a right triangle … ."

RT-4.   In RT-3 you were asked to detect within in a right triangle a relationship between the squares of the lengths of the legs and the square of the length of the hypotenuse. With your group, write a **word equation** (that is, an equation in words) to describe the relationship you noticed. Be prepared to present your group's description to the class.

RT-5.   **The Pythagoarean Relationship**
    a)   The relationship you recorded in the previous problem is often referred to as the Pythagorean relationship, or the Pythagorean theorem. You might have worked with the Pythagorean theorem before in a different form. In this course it will be helpful to remember it as you wrote it in RT-4, as a relationship between the squares of the lengths of the legs and the square of the length of the hypotenuse of a right triangle. Add your word equation for the Pythagorean relationship to your Tool Kit, starting with

"In a right triangle … ."

    b)   The side lengths for the triangles in RT-3 were all integers. Do you think the Pythagorean relationship works for *every* right triangle, including those whose side lengths are not integers? Explain.

RT-6.   Sketch a right triangle.

    a)   Write a sentence or two to explain to a friend how to decide which side is the hypotenuse and which sides are the legs.

    b)   Now share your response to part (a) with your group. If all the responses to part (a) used the same idea, discuss another way to tell the hypotenuse from the legs and write it down. If more than one approach was used, write down all of them.

RT-7.   Solve for each unknown side. Check each result in your "word equation" for the Pythagorean relationship.

a)

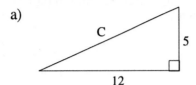

b)

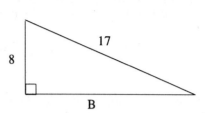

c)   d)

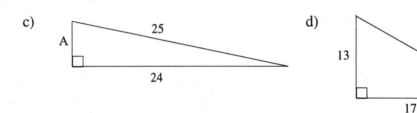

RT-8.   A 10 foot ladder is leaning against a wall. The foot of the ladder is 3.5 feet away from the wall.

a)   Draw a diagram and label the ladder length and its distance from the wall.

b)   How high on the wall does the ladder touch? Write an equation and solve it.

RT-9.   Does $3x^2 = (3x)^2$? That is, does "three times the square of x" equal the "square of three times x ?" Draw a picture to justify your answer.

RT-10.   a)   Add: $\dfrac{5}{8} + \dfrac{1}{3}$.

         b)   Describe what you did in part (a) to find the sum of the two fractions.

         c)   Now find the sum of this pair of fractions: $\dfrac{5}{x} + \dfrac{2}{3}$.

RT-11.   Use the substitution method to solve each systetm of equations for x and y.

         a) $y = 7 - 2x$ b) $y = 7 - 2x$
               $y = 6x - 1$ $3x - y = 3$

RT-12.   a)   Make a table and draw the graph of $y = x^2 - 4$. Mark and label the points where the graph crosses the x-axis.

   b)   Substitute $y = 0$ in the equation $y = x^2 - 4$ to get $0 = x^2 - 4$. Then factor the expression $x^2 - 4$ by thinking of it as $x^2 + 0x - 4$.

   c)   Use the zero product property to find the solutions of the equation $0 = x^2 - 4$ and explain how the solutions are related to the graph.

   d)   Use another method to solve $x^2 - 4 = 0$. Did you get the same results?

RT-13.   Evaluate each of the following expressions for $x = 3$, and then for $x = -2$.

   a) $2x^2$ b) $-5x^2$

   c) $(2x)^2$ d) $3x^3$

   e) $(3x)^3$ f) $(-5x)^2$

RT-14.   Find each of the following products. Generic rectangles may be helpful.

   a) $(x - 3)(x + 5)$ b) $(3x + 1)(x + 4)$

   c) $(2x - 7)^2$ d) $(x - 2)(x^2 + 2x + 1)$

   e) $(2x + 5)(7 - x)$ f) $(3 + 2x)(4 - 3x)$

RT-15.   Factor each of the polynomials in parts (a) through (d). Solve each equation in parts (e) and (f).

   a) $b^2 - 4b + 4$ b) $b^2 - 5b + 4$

   c) $2x^2 - 3x + 1$ d) $2x^2 - 3xy + y^2$

                                                  HINT:

   e) $b^2 - 5b + 4 = 0$ f) $2x^2 - 3x + 1 = 0$

7.2 MAKING SENSE OF SQUARE ROOTS

RT-16.   Karla drew a right triangle with legs of length "1." Because she didn't know the length of the hypotenuse, she labeled it "x_1." On the hypotenuse of the first triangle, Karla then drew another right triangle: the leg lying on the hypotenuse of the first triangle had length "x_1" and the second leg had length "1." Because she didn't know the length of the hypotenuse of the second triangle, Karla labeled it "x_2." She continued in this manner, drawing a right triangle with a leg of length "1" on the hypotenuse of another right triangle to produce this spiral of right triangles:

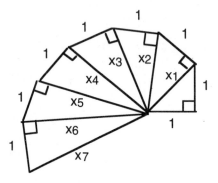

a)   After admiring her drawing, Karla became curious to know the length of the hypotenuse of each triangle, so she started to calculate. Here's what she wrote:

$$1^2 + 1^2 = (x_1)^2, \text{ or } 1 + 1 = (x_1)^2$$
$$\text{so } (x_1)^2 = 2$$
$$x_1 = \sqrt{2}$$

Follow Karla's example and find the lengths x_2, x_3, x_4, etc. Write each length as a square root.

b)   Karla realized that the number "$\sqrt{2}$" didn't mean much to her, so she decided to make a number line to see where $\sqrt{2}$ fit in. To mark units off on the number line, she used the length "1" from the triangles she had drawn, and marked off integers from –5 to 5. She then measured off the segment $x_1 = \sqrt{2}$ on a strip of paper, put one end at "0" on the number line and saw that the other end of the segment fell to the right of "1."

Use Karla's idea and make a number line by following what she did. Then use your solutions from part (a) to mark $\sqrt{2}$, $\sqrt{3}$, $\sqrt{5}$, $-\sqrt{2}$, $-\sqrt{3}$ and $-\sqrt{5}$ on your number line.

c)   Which is greater, $-\sqrt{10}$ or $-\sqrt{12}$? Describe how you decided.

RT-17.   Make a table like the one below on your paper. Find the exact value for each of the unknown values of x, then find a decimal approximation to the nearest tenth.

x^2	exact value of x	approximate value of x
1		
2		
3		
4		
5		
6		
7		
8		
9		
10		

Problems RT-18 and RT-19 involve right triangles. Drawing diagrams will help you visualize and solve them.

RT-18.   A careless construction worker drove a forklift into a telephone pole. The pole cracked (but was not severed) seven feet above the ground. The remainder of the pole fell as if hinged at the crack. The tip of the pole hit the ground 24 feet from its base. If an additional five feet of the pole extends into the ground to anchor it, how long should the replacement pole be?

RT-19.   The four baselines of a baseball diamond form a square, with the bases placed 90 feet apart at the corners. How far is it from home plate directly to second base? Sketch and label a diagram. Use it to write an equation, then solve the equation.

RT-20.   Write each list of numbers in order from smallest to largest:

a)   2, $\sqrt{5}$, 1.5

b)   3, $\sqrt{2}$, 1.6

c) $\sqrt{7}$, 2, $\frac{5}{2}$

d)   10, $\sqrt{102}$, 10.8

e) $\sqrt{121}$, 10.9, 11.1

f) $-\sqrt{3}$, −1.62, −1.83

RT-21. **Fraction Busters** In this course you've been solving equations that contain algebraic fractions. Some of the equations took several steps -- applications of the idea of "un-doing" and "keeping the balance" -- to solve. Here's a more efficient approach, based on the same ideas, where the "un-doing" is accomplished by multiplying by a "fraction buster:"

Suppose you want to solvethe equation $\frac{x}{3} + \frac{x}{5} = 2$.

The hard part of this problem is dealing with the fractions. You could add them by first writing them in terms of a common denominator, ...*but there is an easier way.*

$$\frac{x}{3} + \frac{x}{5} = 2$$

You can avoid dealing with fractions by "eliminating" the denominators. To do this, you'll still need to **find a common denominator, and** then you'll **multiply both sides of the equation by that common denominator**.
In this case the lowest common denominator is 15, and you could multiply both sides of the equation by 15:

The lowest common denominator of $\frac{x}{3}$ and $\frac{x}{5}$ is 15.

$$15 \cdot (\frac{x}{3} + \frac{x}{5}) = 15 \cdot 2$$

The number you use to eliminate the denominators is called a **fraction buster**. Here we used **15** to bust the fractions.

$$15 \cdot \frac{x}{3} + 15 \cdot \frac{x}{5} = 15 \cdot 2$$

The result is a valid equation *without fractions*: It looks like many you have seen before and can be solved it in the usual way.

$$5x + 3x = 30$$
$$8x = 30$$
$$x = \frac{30}{8} = \frac{15}{4} = 3.75$$

You will, of course, always want to check the answer!

Check: $\frac{3.75}{3} + \frac{3.75}{5} = 2$
$$1.25 + 0.75 = 2$$

Copy the following problem on your paper. Fill in each of the lines labeled (a) through (e) to explain how the equation to its right was obtained from the equation above it.

Solve the equation:

$$\frac{4}{x} + \frac{3}{2x} = \frac{11}{6}$$

a) _____ $6x(\frac{4}{x} + \frac{3}{2x}) = 6x \cdot (\frac{11}{6})$

b) _____ $6x \cdot (\frac{4}{x}) + 6x \cdot (\frac{3}{2x}) = 6x \cdot (\frac{11}{6})$

c) _____ $24 + 9 = 11x$

d) _____ $33 = 11x$

e) _____ $3 = x$

RT-22. Solve each of the following equations. In each case, find the lowest common denominator first, as in problem RT-21.

a) $\dfrac{x}{2} + \dfrac{x}{3} = \dfrac{1}{6}$

b) $x + \dfrac{x}{2} + \dfrac{x}{3} = 22$

c) $\dfrac{4}{x} - 1 = 7$

d) $\dfrac{1}{x} + \dfrac{1}{2x} = 3$

RT-23. Find each sum or difference of fractions:

a) $\dfrac{4}{3} - \dfrac{2}{7}$

b) $\dfrac{5}{2x} - \dfrac{6}{x}$

c) $\dfrac{3}{x} + \dfrac{1}{2} + \dfrac{5}{2x}$

RT-24. For each right triangle, use the given information to find the length of the missing side. Write your lengths in both decimal and square root ($\sqrt{}$) form.

Triangle	leg#1	leg#2	hypotenuse
a)	4	7	
b)	13	19	
c)	49.77	14	
d)	3		7
e)	6	6	
f)	33.24	39.6	

RT-25. Copy the diagram and then find x to the nearest tenth:

3x   17

x

RT-26.   **Less is More?** Janelle has $40 in her "spare change" jar and is adding to it at the rate of $7 per week. Jeanne has $125 hidden away and is putting aside extra cash at a rate of $3 per week. After how many weeks will both women have the same amount of money set aside?

    a)   Make a graph to solve the problem.

    b)   Write equations to solve the problem algebraically.

    c)   If you were presenting your solution to students who didn't know the problem, would the graph help them visualize the problem? Explain your answer.

RT-27.   Solve each of the following systems for x and y.

    a) $y = 4x$
        $x + y = -1$

    b) $-2x + y = 3$
        $x = 3y$

RT-28.   To find the product $2(x + 1)(2x + 3)$ we could first multiply 2 and $(x + 1)$, and then multiply the result by $2x + 3$, <u>OR</u> we could first multiply out $(x + 1)(2x + 3)$, and then multiply that result by 2. Either way, finding the product $2(x + 1)(2x + 3)$ is a two-step operation.

Find each of the following products.

    a) $3(x - 1)(5x - 7)$

    b) $5(x + 3)(2x - 5)$

    c) $xy(x - 3y)(x + 2y)$

    d) $2x^2(x - 7)(x + 7)$

RT-29.   Write one polynomial to represent each of the following sums and differences.

    a) $(8x^2 + 5x + 7) - (3x^2 + 2x + 2)$

    b) $(8x^2 + 5x + 7) - (3x^2 - 2x + 2)$

    c) $(5x^2 + 14x + 3) - (2x^2 - 9x + 5)$

    d) $(16x^2 + 3x - 8) - (7x^2 + 8x - 10)$

    e) $(10x^2 - 13x + 1) - (3x^2 - 10x + 1)$

RT-30.   A jar with a lot of coins in it is $\frac{1}{3}$ pennies, $\frac{1}{3}$ nickels, and $\frac{1}{3}$ dimes. When Juan puts 24 more pennies in the jar the probability of pulling out a penny becomes $\frac{1}{2}$. How many pennies, nickels and dimes were in the jar to begin with?

7.3 RIGHT TRIANGLES AND DIAGRAMS

Problems RT-31 and RT-32 involve right triangles. Making models will help you draw diagrams that will be useful in solving the problems.

RT-31.   The Flat family's roof is 32 feet wide and 60 feet long. Their television antenna, which rises 30 feet above the center of the roof, is anchored by wires that are attached five feet below the top of the antenna. The wires are attached to the house at each corner of the roof and at the midpoint of each edge of the roof. How long must each wire be?

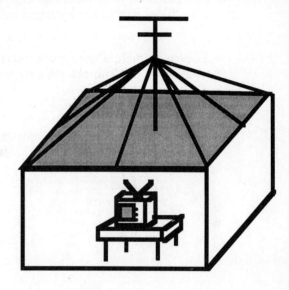

RT-32.   Jay wants to ship a fishing pole to his summer home in Washington. The longest section of the pole is 40 inches. Jay can only find a rectangular box that has dimensions 24 inches by 30 inches by 18 inches. If the 40-inch section of pole is placed corner-to-corner inside the box, will it fit? Show how you know.

RT-33.   **Area and the Pythagorean Theorem** The following problem gives a way to visualize the Pythagorean theorem.

   a)   On dot paper, draw a right triangle with legs six units and eight units long, respectively. Label the sides with their lengths.

   b)   Use the Pythagorean theorem to determine the length of the hypotenuse of your triangle.

   c)   Using dot paper units, measure the hypotenuse to confirm your calculation in part (b). Label the hypotenuse with its length.

   d)   Using the shorter leg as one of the sides, draw a square with sides six units long. Repeat this step with the longer leg (eight units long). Your drawing should look something like this:

   e)   Using the hypotenuse as one of the sides, draw another square.

[PROBLEM CONTINUED ON NEXT PAGE]

RT-33. continued

      f)   Inside each of the three squares, write its area.

      g)   Write a word equation describing how the areas of the squares on the legs relate to the area of the square on the hypotenuse.

      h)   How is your word equation in part (g) related to the word equation you wrote in your Tool Kit for the Pythagorean relationship?

RT-34.   For each graph, draw a right triangle and determine the length of the given line segment. Do your work on your resource page for RT-34.

a)
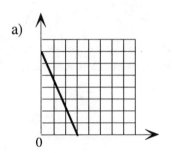

b)
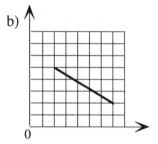

c)
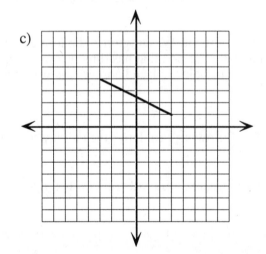

d)
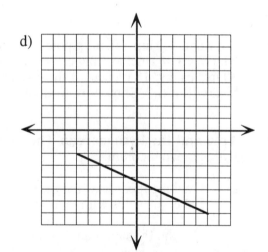

RT-35.   Solve each of the following equations for x.

    a) $\frac{x}{2} + \frac{x}{3} = 5$

    b) $\frac{5}{x} - 8 = 12$

    c) $2x + y = 29$

    d) $\frac{5 + x}{7} + \frac{12x}{18} = 5$

    e) $x^2 + 6x - 7 = 0$

RT-36. Draw coordinate axes on a piece of graph paper. Start at the point (10, 6). Slide eight units to the left, then seven units down, and stop.

   a) What are the coordinates of the stopping point?

   b) What is the distance between the starting and stopping points?

RT-37. Washek claims the y-intercept of the graph of $2x - y = 5$ is 5. Lida says that it is –5. Which claim is correct? Explain how you know.

RT-38. Solve for x: $\dfrac{4}{x} + \dfrac{3}{5} = \dfrac{7}{10}$.

RT-39. Max paid $2379.29 for a computer that was originally priced at $2500.00. What percent of the original price did Max pay? Write an equation and solve it.

RT-40. Solve each system of equations for x and y:

   a) $y = 0$

   $y = x^2 + 5x + 4$

   b) $y = \dfrac{-2}{3}x + 4$

   $\dfrac{1}{3}x - y = 2$

RT-41. Factor each of the following polynomials.

   a) $y^2 - 3y - 10$

   b) $x^2 + 27x + 50$

   c) $2x^2 - 7x + 3$

   d) $5x^2 + 6xy + y^2$

RT-42. Without using your calculator, put each of these lists of numbers in order from smallest to largest. (If you need to use your calculator, use it as little as possible.)

   a) $3\sqrt{2}$, 4, 7.5, $2\sqrt{3}$

   b) $10\sqrt{3}$, $3\sqrt{10}$, 12, 8

   c) –5, –8.3, $4\sqrt{2}$, $-2\sqrt{6}$

   d) $3\sqrt{17}$, $\sqrt{100}$, 11.5, 10.9

7.4 CALCULATING WITH SQUARE ROOTS

RT-43.   You know that $x \cdot x = x^2$.

a)   Calculate $(\sqrt{3})^2$. Now calculate $\sqrt{3^2}$. What do you notice?

b)   Calculate $(\sqrt{4.7})^2$. Now calculate $\sqrt{4.7^2}$. What do you notice?

c)   Look for a pattern in parts (a) and (b), and then try it on another number. Describe your results.

RT-44.   Predict whether each statement is true or false. Then use your calculator to check.

a) $\sqrt{2} + \sqrt{3} = \sqrt{5}$

b) $\dfrac{\sqrt{12}}{\sqrt{4}} = \sqrt{3}$

c) $(\sqrt{2})(\sqrt{3}) = \sqrt{6}$

d) $\dfrac{\sqrt{7}}{\sqrt{2}} = \sqrt{3.5}$

e) $\sqrt{5} - \sqrt{2} = \sqrt{3}$

f) $\sqrt{16} + \sqrt{3} = \sqrt{19}$

g) $\dfrac{\sqrt{6}}{\sqrt{3}} = \sqrt{2}$

h) $(\sqrt{5})(\sqrt{7}) = \sqrt{35}$

RT-45.   With your group try to generalize what you observed in the previous problem. Write several general statements about adding, subtracting, multiplying, and dividing **unlike** square roots. For each of your statements, write three examples that support it.

RT-46.   Rewrite each of following expressions using as few square roots symbols and numbers as possible. Check your answers with a calculator.

a) $(\sqrt{3})(\sqrt{7})$

b) $\dfrac{\sqrt{20}}{\sqrt{5}}$

c) $\sqrt{3} + \sqrt{7}$

d) $\dfrac{\sqrt{144}}{\sqrt{3} \cdot \sqrt{4}}$

e) $\sqrt{7} - \sqrt{3}$

f) $\dfrac{\sqrt{21}}{\sqrt{3}}$

g) $\sqrt{20} + \sqrt{5}$

h) $\dfrac{1}{\sqrt{4}}$

RT-47.Multiply the following pairs of binomials.

a)$(2x + 4)(3x + 5)$

b)$(5x - 2)(x + 3)$

c)$(3x - 1)(2x - 5)$

d)$(5x + 4)(x - 2)$

e)In multiplying binomials, such as $(3x - 2)(4x + 5)$, you might use a generic composite rectangle, or you could use a shortcut known as F.O.I.L. (First, Outside, Inside, Last).

F.O.I.L. is an acronym which reminds you of a shortcut for multiplying binomials. For the product $(3x - 2)(4x + 5)$ we have

F.multiply the FIRST terms of each binomial$(3x)(4x) = 12x^2$
O.multiply the OUTSIDE terms$(3x)(5) = 15x$
I.multiply the INSIDE terms$(-2)(4x) = -8x$
L.multiply the LAST terms of each binomial$(-2)(5) = -10$

Finally, we compute $12x^2 + 15x - 8x - 10 = 12x^2 + 7x - 10$.

Notice how the generic composite rectangle relates to the F.O.I.L. method:

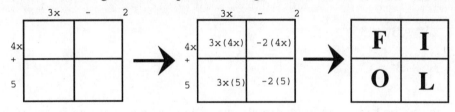

RT-48.Use any method you choose to multiply these binomials:

a)$(7x - 4)(3x + 2)$

b)$(9x + 7)(4x - 3)$

c)$(2x - 5)(3x - 10)$

d)$(5x - 4)(3x - 2)$

RT-49.Determine which of the following statements are true and which are false.

a)$\sqrt{25} + \sqrt{9} = \sqrt{34}$

b)$\sqrt{8} - \sqrt{3} = \sqrt{5}$

c)$(\sqrt{1.5})(\sqrt{1.5}) = \sqrt{2.25}$

d)$\sqrt{9} + \sqrt{8} = \sqrt{17}$

e)$\sqrt{100} - \sqrt{36} = \sqrt{64}$

f)$(\sqrt{5})(\sqrt{3}) = \sqrt{15}$

g)$\dfrac{\sqrt{90}}{\sqrt{3}} = \sqrt{30}$

h)$(\sqrt{3704})^2 = 3704$

[PROBLEM CONTINUED ON NEXT PAGE]

RT-49. continued

    i) $\sqrt{512^2} = 512$ j) $\sqrt{5^2 \cdot 2} = 5\sqrt{2}$

    k) $\sqrt{3 \cdot 3 \cdot 7} = 3\sqrt{7}$ l) $(\sqrt{25})(\sqrt{9}) = \sqrt{225}$

    m)   Do your answers in this problem support your statements in RT-45 ? If not, describe the conflict. Did you discover anything new?

RT-50.   Graph the equation $y = \sqrt{x}$, and find the domain (the possible x-values). Are there any values for x which cannot work in this rule? Explain.

RT-51.   Solve each of the following equations for A.

    a) $A^2 = 152 - 122$

    b) $A^2 + (26.4)^2 = 292$

    Solve each of the following equations for B.

    c) $122 - 52 = B^2$

    d) $25^2 + B^2 = 35^2$

    e) $A^2 + B^2 = C^2$

RT-52.   For each of the following polynomials, write the greatest common factor of the terms.

    a) $2x^2 + 2x$ b) $27x^2 - 3$

    c) $12x^3y - 3xy^3$ d) $5x^2y - 30xy + 45y$

RT-53.   Solve each of the following equations.

    a) $\dfrac{x}{3} = \dfrac{4}{5}$ c) $\dfrac{x}{x + 1} = \dfrac{5}{7}$

    b) $\dfrac{3}{x} = \dfrac{4}{5}$ d) $\dfrac{2}{y} = \dfrac{3}{y + 5}$

RT-54.   In the diagram below, the two triangles are similar. Find x.

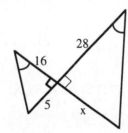

7.5 EQUIVALENT SQUARE ROOT EXPRESSIONS

RT-55.   **Simplifying Square Roots** Before calculators were universally available, if people wanted to calculate approximate values for numbers like $\sqrt{45}$ they could use long tables of square root values or use Guess and Check over and over, depending on how accurate an answer they wanted. They also found that if they memorized the square roots of the integers from 1 to 10, and used factoring, then they could figure out the square roots of many larger numbers fairly quickly. In factoring, what they looked for were factors that were perfect squares.

Here's the method they developed for "simplifying" square roots based on factoring out perfect squares:

Example Simplify $\sqrt{45}$.

You want to rewrite $\sqrt{45}$ in an equivalent form.

First factor 45 so that one of the factors is a **perfect square**:
$$\sqrt{45} = \sqrt{\mathbf{9 \cdot} 5}$$
$$= \sqrt{\mathbf{9}} \cdot \sqrt{5}$$
Then rewrite the square root of the perfect square:
$$= 3\sqrt{5}$$

With your calculator you can verify that $\sqrt{45} \approx 6.71$ and $3\sqrt{5} \approx 3(2.236) \approx 6.71$, so both forms, $\sqrt{45}$ and $3\sqrt{5}$, are equivalent.

Here are two more examples:
$$\sqrt{27} = \sqrt{\mathbf{9}} \cdot \sqrt{3} = 3\sqrt{3}$$
$$\sqrt{72} = \sqrt{\mathbf{36}} \cdot \sqrt{2} = 6\sqrt{2}$$

Note that we chose to write $\sqrt{72}$ as $\sqrt{\mathbf{36}} \cdot \sqrt{2}$, rather than $\sqrt{\mathbf{9}} \cdot \sqrt{8}$ or $\sqrt{4} \cdot \sqrt{18}$, because 36 is the largest perfect square factor of 72. However, since
$$\sqrt{4} \cdot \sqrt{18} = 2 \cdot \sqrt{18} = 2 \cdot \sqrt{\mathbf{9 \cdot} 2} = 2 \cdot \sqrt{\mathbf{9}} \cdot \sqrt{2} = 2 \cdot 3 \cdot \sqrt{2} = 6\sqrt{2},$$
we still get the same answer if we simplify $\sqrt{4} \cdot \sqrt{18}$ in several steps. Try it with $\sqrt{9} \cdot \sqrt{8}$.

Now write each of the following square roots using the smallest numbers possible.

a) $\sqrt{50}$ b) $\sqrt{18}$

c) $\sqrt{48}$ d) $\sqrt{24}$

e) $\sqrt{250}$ f) $\sqrt{1000}$

RT-56.   Match the number in the left-hand column with an equivalent number in the right-hand column. Whenever possible do this without using a calculator.

a) $2\sqrt{15}$ $\sqrt{96}$

b) $3\sqrt{2}$ $\sqrt{12}$

c) $3\sqrt{5}$ $\sqrt{60}$

d) $3\sqrt{6}$ $\sqrt{72}$

e) $5\sqrt{2}$ $\sqrt{24}$

f) $2\sqrt{3}$ $\sqrt{54}$

g) $4\sqrt{6}$ $\sqrt{28}$

h) $2\sqrt{5}$ $\sqrt{8}$

i) $2\sqrt{6}$ $\sqrt{50}$

j) $6\sqrt{2}$ $\sqrt{45}$

k) $4\sqrt{3}$ $\sqrt{20}$

l) $5\sqrt{3}$ $\sqrt{75}$

m) $2\sqrt{7}$ $\sqrt{48}$

n) $2\sqrt{2}$ $\sqrt{18}$

RT-57.   Graph each pair of points and find the distance between them.

a) $(3, -6)$ and $(-2, 5)$

b) $(5, -8)$ and $(-3, 1)$

c) $(0, 5)$ and $(5, 0)$

d)   Write the distance you found in part (c) in simplified square root form.

RT-58.   Start with the generic point (x, y). Imagine sliding the point six units to the left and 4.5 units up.

a)   What are the coordinates of the point after the slide?

b)   Draw a diagram. What is the length of the segment that joins the original point and its image after the slide?

RT-59.   Draw a diagram and use it to find the distance between the points (−4, 7) and (29, 76).

RT-60.   A young redwood tree in Muir Woods casts a shadow 12 feet long. At the same time, a five-foot tall tourist casts a shadow two feet long. How tall is the tree? Draw a diagram, use ratios to write an equation, and then solve the equation.

RT-61.   A Mercedes is traveling east from Elko at 50 miles per hour and a Yugo is traveling west from Elko at 60 miles per hour. Both cars started at the same time.

    a)   Draw a diagram showing the cars' positions.

    How far apart are the cars ...

    b)   after one hour?

    c)   after two hours?

    d)   after three hours?

    e)   after x hours?

    f)   In your drawing label the distance each car travels in x hours. Write an equation to show how far apart the cars are after x hours of travel.

RT-62.   One car is traveling due east and another car is traveling due north. If they both started from the same point at the same time going 40 miles per hour, how far apart are the cars ...

    a)   after one hour?

    b)   after two hours?

    c)   after three hours?

    d)   after n hours?

RT-63.   Two trucks leave a rest stop at the same time. One heads due east; the other heads due north and travels twice as fast as the first truck. The trucks lose radio contact when they are 47 miles apart. (They are obviously using an illegal power amplifier since FCC regulations limit CB's to five watts which have a normal range of about five miles.) How far has each truck traveled when they lose contact?

RT-64. **More examples of factoring** One way to factor a polynomial is first to look for common
factors:

	Example 1	Example 2
original polynomial	$2x^2 - 8$	$4x^3y - xy^3$
(**common factors**)·(remaining polynomial)	$2·(x^2 - 4)$	$xy·(4x^2 - y^2)$

And then, if possible, continue to factor. Here both $x^2 - 4$ and $4x^2 - y^2$ are **differences of
two squared terms**:
$$x^2 - 4 = (x + 2)(x - 2) \quad \text{and} \quad 4x^2 - y^2 = (2x + y)(2x - y).$$
So when we factor $2x^2 - 8$ we get
$$2x^2 - 8 = 2(x^2 - 4) = 2(x + 2)(x - 2),$$
and when we factor $4x^3y - xy^3$ we get
$$4x^3y - xy^3 = xy(4x^2 - y^2) = xy(2x + y)(2x - y).$$

Factor each of the following expressions as completely as possible.

a) $27x^2 - 3$

b) $12x^3y - 3xy^3$

c) $5x^2y - 30xy + 45y$

RT-65. Two trucks leave the same rest stop at the same time traveling at the same speed. One heads
south, and the other goes west. When the two trucks lose CB contact, they are 53 miles
apart. How far has each traveled?

RT-66. Factor each of the following expressions.

a) $x^2 - 25$

b) $4x^2 - 49$

c) $16y^2 - 1$

d) $100x^2 - 49y^2$

e)   What is the same about the expressions in parts (a) through (d) ? Write a general
description for them.

The expressions in parts (f) through (i) are in some ways similar to, and in some ways
different from, those in parts (a) through (d). Some of them can be factored and some cannot.
Factor any that can be factored, and describe how each one differs from your general
description in part (e).

f) $x^2 + 36$

g) $8 - 64y^2$

h) $2x^2 - 25$

i) $9y^2 + 36x^2$

RT-67.   Start at the point $(-4, 5)$. Slide six units to the right and 10 units down. Write the coordinates of the resulting point and find the distance between it and the point $(-4, 5)$.

RT-68.   Using what you learned in RT-34 and RT-57, find the distance between each pair of points. Write the distances you calculate in parts (a) and (b) in square root form, simplified square root form, and decimal form rounded to the nearest 0.01. Write the distances you calculate in parts (c) and (d) only in decimal form to the nearest 0.01.

   a) $(0, 0)$ and $(4, 4)$

   b) $(-2, 4)$ and $(4, 7)$

   c) $(12, 18)$ and $(-16, -19)$

   d) $(0, 0)$ and $(25, 25)$

RT-69.   Factor each polynomial completely. Look for a common factor first.

   a) $2xy - 4y^2$

   b) $3x^2 - 6x + 3$

RT-70.   Matilda and Nancy are 60 miles apart, bicycling toward each other on the same road. Matilda rides 12 miles per hour and Nancy rides eight miles per hour. In how many hours will they meet? Draw a diagram, then write an equation and solve it.

RT-71.   Two cars leave a parking lot at the same time. One goes south and one goes west. One is traveling at 55 miles per hour and the other at 45 miles per hour. How long will it take before the cars are 150 miles apart?

RT-72.   Solve each of the following systems for x and y.

   a) $y = 3x - 5$
         $y = 5x - 9$

   b)   If $y = 3x - 5$ and $y = 5x - 9$ were graphed on the same set of axes, at what point would the lines intersect?

   c) $y = 2x + 5$
         $3x + 2y = 31$

   d) $x + 2y = 1$
         $3x - 2y = -5$

RT-73.   Solve for x: $\dfrac{4}{3x} + \dfrac{6}{x} = 9$

7.6 MORE DIAGRAMS TO EQUATIONS

RT-74. **The Buckled Railroad Track** A new railroad line was installed in the Central Valley. In order to reduce derailments along a two-mile stretch, the track was made with straight one-mile long rails. The rails were laid in the winter and they expanded in the heat of the following summer. Indeed, each mile-long rail expanded one foot in length! Ordinarily, because the rails do not bend, they would jut to the side. However, in this strange case, the rails jutted upward where their ends met. How high above the ground were the expanded rails at the joint?

Let H be the height (in feet) of the tracks above the ground where two rails come together.

a) **Make a guess** of how large you think H might be: big enough for you to stick your arm between the ground and the tracks? big enough walk through? Could you drive a car under the buckled tracks?

b) Draw a diagram and label it. Use your picture to help calculate H. Note that a mile is 5280 feet long.

c) How does the value you calculated for H compare to your guess in part (a) ?

RT-75. A train leaves Roseville at 6:00 am heading east at 45 miles per hour. A different train leaves the same station heading west at 7:00 am at 45 miles per hour also. What time are the two trains 240 miles apart? If you need help getting started, draw a diagram.

RT-76. Trevor rode his bike to Folsom Lake at 20 kilometers per hour. After going for a swim, he found that the back bicycle tire was flat, so he had to walk home. Trevor walked at eight kilometers per hour all the way home. If his round trip traveling time was seven hours, how far was it to the lake? If you need help getting started, draw a diagram.

RT-77. Janis is going to fence off a rectangular garden. She will use an existing wall along the back, and she wants the length to be twice as long as the width. The total amount of fencing material she has is 84 meters long. What are the width and length of her garden? (Be sure to draw a diagram and write an equation.)

RT-78. The lengths of the congruent sides of an isosceles triangle are $2x + 5$ and $3x - 8$ units, respectively. Find the lengths of each of the congruent sides.

RT-79. a) Use your calculator to graph $y = \sqrt{x + 2}$ for $-2 \le x \le 6$.

b) What happens if you use $x = -5$ in part (a) ?

RT-80. Use your calculator to graph $y = \sqrt{x - 5}$. Before you graph, try $x = 4$ and $x = 5$.

RT-81.   a)   Use your calculator to graph $y = \sqrt{2x}$. What is the domain (the possible x-values) for the equation?

b)   Write one or two sentences to compare your graph of $y = \sqrt{2x}$ to your graph of $y = \sqrt{x}$ in RT-50 . Describe how the two graphs are alike and how they are different.

RT-82.   a)   Use your calculator to graph $y = \sqrt{x} + 2$. What is the domain (the possible x-values) for the equation?

b)   Write one or two sentences to compare your graph of $y = \sqrt{x} + 2$ to your graph of $y = \sqrt{x}$ in RT-50. Describe how the two graphs are alike and how they are different.

RT-83.   Cassie walked for a while at three miles per hour and then continued her journey on a bus at 15 miles per hour. Her time on the bus was twice as long as her time walking. How long did she ride on the bus if the total distance she covered was 66 miles? If you need help getting started, draw a diagram.

RT-84.   Factor and solve each of the following quadratic equations.

a) $y^2 + 6y - 16 = 0$

b) $x^2 + 8x + 16 = 0$

c) $5z^2 - 13z = 0$ Look for a common factor.

d) $x^2 - 13x + 40 = 0$

e) $2x^2 + 15x + 7 = 0$

RT-85.   Solve each system of equations for x and y:

a) $3x + y = 9$
$2x + y = 1$

b) $x = 3y - 7$
$3x - y = 3$

c)   Where would the graphs for part (b) intersect?

RT-86.   Copy and complete the table for (x, y), then write the rule.

x	−3	−2	−1	5		1	$\frac{1}{3}$	6
y	−10	−7		14	−1			17

RT-87.   Solve each of the following equations for x.

a) $\dfrac{x-3}{5} = 12(x-1)$

b) $\dfrac{10}{x} + \dfrac{10}{2x} = 10$

c) $2x^2 - 6x = 0$

d) $\dfrac{5+x}{6} = \dfrac{3x}{14}$

RT-88.   **Chapter 7 Summary: Rough Draft** Some of the problems you did Chapter 7 focused on developing four focal ideas of the chapter and others reinforced concepts and skills introduced in previous chapters. Re-read the chapter's Introduction and then look back through the chapter to find where each main idea was developed

Write your answers to the following questions in rough draft form **on separate sheets of paper**, and be ready to discuss them with your group at the next class meeting. Focus on the **content**, not neatness or appearance, as you write your summary draft. You will have the chance to revise your work after discussing the rough draft with your group.

a)   In this chapter you've used diagrams to help write equations. From your work, select two problems that best show your understanding of how diagrams can be useful tools for solving problems. Write complete sentences to describe how you did each of the selected problems. Then tell why you chose the problems that you did.

b)   Show how fraction busters can be used to solve the equation $\dfrac{5}{6x} + \dfrac{8}{15} = \dfrac{2}{3}$.

c)   Select a problem you didn't understand before, but now know how to do. Show all your work and a complete solution. Explain why you chose the problem.

7.7 SUMMARY AND REVIEW

RT-89.   **Chapter 7 Summary: Group Discussion** Take out the rough draft summary you completed in RT-88.

Take turns to describe the problems group members chose to illustrate their understanding of how word problems, diagrams, and equations are related. Each group member should:

- explain the problems he or she chose to illustrate the main ideas;

- explain why he or she chose those particular problems; and

- explain a problem he or she didn't understand before, but now can solve, and explain why the problem was chosen.

This is your chance to make sure your summary is complete, update your Tool Kits, and work together on problems you may not be able to solve yet. Use homework time to revise your summary as needed.

RT-90.   Graph $y = \sqrt{11 - x}$, and find its domain (the possible x-values).

RT-91.   Use your calculator to graph $y = \dfrac{x}{\sqrt{x}}$, and find the domain (the possible x-values).

RT-92.   Compare your graphs from RT-50, 79, 80, 81, 82, 90, and 91. Write a short paragraph to describe several of your observations about square root graphs and their equations.

Solve each of the following problems by drawing a diagram, and then writing and solving an equation.

RT-93.   How long must a wire be to reach from the top of a 13-meter telephone pole to a point on the ground nine meters from the foot of the pole?

RT-94.   A 10-meter ladder is leaning against a building. The bottom of the ladder is five meters from the building. How high is the top of the ladder?

RT-95.   Cleopatra rode an elephant to the outskirts of Rome at two kilometers per hour and then took a chariot back to camp at 10 kilometers per hour. If the total traveling time was 18 hours, how far was it from camp to the outskirts of Rome?

RT-96.   Find the missing side lengths in each triangle. Write each length in simplified square root and decimal forms.

a)   b)   c)

RT-97.   Determine whether each of the following statements is true or false.

a) $\sqrt{12} = \sqrt{4} \cdot \sqrt{3}$ b) $\sqrt{48} = 4\sqrt{3}$

c) $\sqrt{100} - \sqrt{64} = \sqrt{36}$ d) $\dfrac{\sqrt{10}}{2} = \sqrt{5}$

e) $\sqrt{2} + \sqrt{3} = \sqrt{5}$ f) $\sqrt{7} \cdot \sqrt{7} = \sqrt{49}$

g) $\sqrt{3} \cdot \sqrt{3} = 3$ h) $\sqrt{5} \cdot \sqrt{5} = (\sqrt{5})^2$

RT-98.   Find the distance between each pair of points.

a)   (5, –8) and (–3, 1)

b)   (0, –3) and (0, 5)

RT-99.   Solve each of the following equations for x.

a) $x^2 - 140 = 4$

b) $x(x+1)(x+2) = 0$

c) $1 - \dfrac{5}{6x} = \dfrac{x}{6}$

d) $2x + 3y = 5$

e) $\dfrac{3}{x} - 7 = 10$

f) $\dfrac{1}{x} + \dfrac{2}{3x} = 5$

RT-100.   Solve each system of equations for x and y :

a) $y = 9 - x$
       $y = 9 + x$

b) $3x - 2y = 20$
       $x = 5 - y$

RT-101. Two cars leave San Francisco at the same time going in opposite directions. One travels at 99 kilometers per hour and the other travels at 81 kilometers per hour. In how many hours will they be 300 kilometers apart?

RT-102. Solve each equation by factoring.

a) $3x^2 - 4x + 1 = 0$

b) $x^2 - 4x = 5$

RT-103. A certain rectangle has area $3x^2 - xy - 2y^2$ square units. Find its dimensions.

RT-104. A rectangular garden of area 63 square meters has a width two meters less than its length. Find the length and width of the garden.

RT-105. In the last student council election at Sierra Nevada College only about 43% of the students voted. A total of 1576 ballots were counted. About how many students attend Sierra Nevada College?

RT-106. Find at least ten points that are 10 units away from the origin and have integer coordinates. (Hint: The point (6, 8) works.)

RT-107. Evaluate each of the following expressions with $x = 2$ and $x = -3$.

a) $2x^2$

b) $(2x)^2$

c) $-5x^2$

d) $(3x)^3$

e) Write an equivalent expression for $(5x)^2$ without parentheses.

RT-108. **Chapter 7 Summary: Revision** Use the ideas your group discussed in RT-89 to revise the your rough draft of a Chapter 7 Summary. Your presentation should be thorough and organized, and should be done on paper separate from your other work.

RT-3. Sides of Right Triangles

How complicated is the relationship among the legs and hypotenuse of a right triangle? To help answer this question, complete the table for the five given right triangles. Look for a pattern in the three columns on the far right. What do you observe?

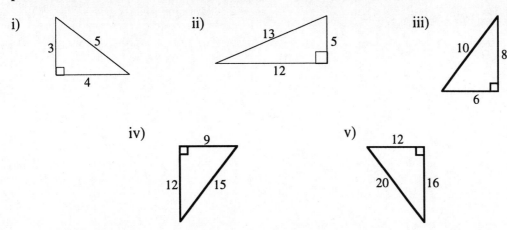

i)

ii)

iii)

iv)

v)

	length of leg#1	length of leg#2	length of hypotenuse	$(\text{length of leg\#1})^2$	$(\text{length of leg\#2})^2$	$(\text{length of hypotenuse})^2$
i)						
ii)						
iii)						
iv)						
v)						

Algebra Tool Kit: the "what to do when you don't remember what to do" kit

Algebra Tool Kit: the "what to do when you don't remember what to do" kit

Dot Paper

Dot Paper

RT-34. For each graph, draw a right triangle and use it to determine the length of the given line segment.

a)
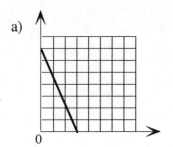

b)
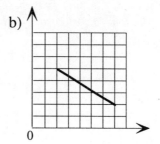

c)
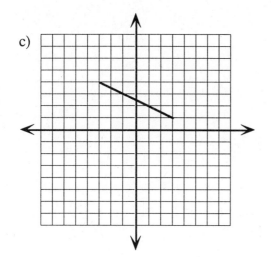

d)
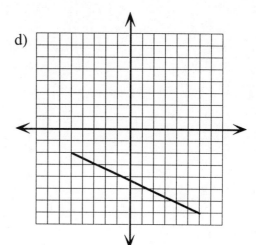

Chapter 8

THE GRAZING GOAT:
Area and
Subproblems

CHAPTER 8

THE GRAZING GOAT:
AREA AND SUBPROBLEMS

This Grazing Goat Problem is an example of a complicated situation that can be most easily attacked by looking for subproblems (smaller, more manageable problems):

> A barn 15 meters by 25 meters stands in the middle of a large grassy field.
> Tied by a rope to one corner of the barn is a hungry goat. Over what area of the
> field can the goat graze if the rope is x meters long?

Near the beginning of this chapter you will find the areas of complex geometric figures by sectioning them into simpler shapes. That is, you will solve a complicated problem by first finding and solving simpler problems. Later, you will use this same **subproblem** approach to "uncomplicate" complicated algebraic processes. To do this, you have to be able to step back from the whole problem and identify useful subproblems before you begin.

The ability to identify subproblems could be one of the most useful skills you learn in this course!

In this chapter, you will have the opportunity to:

- use the strategy of identifying, writing and solving subproblems to solve a larger problem;

- further the development and application of the concepts of area and perimeter for rectangles, circles and triangles;

- use multi-step processes, such as simplifying rational expressions; and

- solve many types of equations using subproblems including equations with radical expressions and systems of equations using elimination.

CHAPTER CONTENTS

8.1 FINDING AREAS USING SUBPROBLEMS

GG-1.   Use dot paper to find the area of each of the following figures. Explain two ways to find each area.

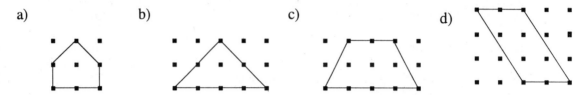

a)   b)   c)   d)

GG-2.   a)   On dot paper make four triangles, each with an area of one square unit, and all in different shapes. Show how you figured out the area of each triangle.

          b)   Repeat part (a) making four different triangles, each with area two square units.

          c)   Repeat part (a) making four different triangles, each with area four square units.

          d)   Write a statement to describe how to find the area of any triangle.

GG-3.   You want to tile your bathroom floor. How many 3" by 3" square tiles will fit on a 6' by 8' rectangular floor? Identify a subproblem you will need to solve first.

GG-4.   Without the use dot paper, find the area of this figure:

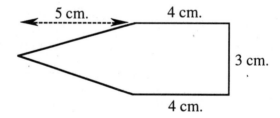

GG-5.   a)   Find the missing dimensions for this figure. (All corners form right angles.)

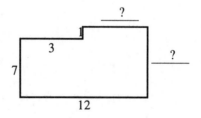

[PROBLEM CONTINUED ON NEXT PAGE]

GG-5. continued

b)   Find the missing dimensions for this figure. (All corners form right angles.)

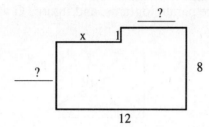

c)   For the figure in part (b), if x = 2, then the perimeter is 40 units and the area is 94 square units. Explain to a pre-algebra student **how to find** the perimeter and area when x is 2.

d)   For the figure in part (b), does the length of the side marked "x" matter when you calculate the perimeter? Do you think the value for "x" affects the area of the figure? Make guesses to assign various values to x and then organize your data in a table. With each guess, record the resulting perimeter and area.

x	Perimeter	Area
1		
2	40 units	94 square units
3		

e)   What is the largest number you could use for x? Why?

f)   What do you notice about the perimeter? Why is this true?

g)   What do you notice about the area?

GG-6.   For each figure below, find its area <u>and</u> its perimeter. In most cases, you will need to solve
         one or more subproblems before you can solve the original problems of finding the area or
         perimeter. In part B, **state all the subproblem(s)** you use to solve each problem.
         Write your answers in square root and decimal form.
         Note: Figures E and F are parallelograms, and figures G and H are trapezoids.

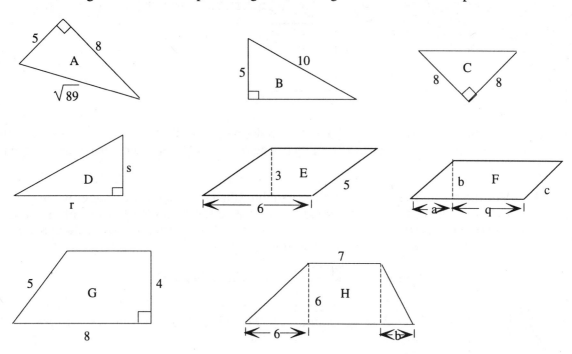

GG-7.   Copy this figure onto dot paper: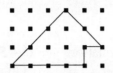

a)   Enlarge the original figure by making its corresponding sides three times as long.

b)   Now enlarge the sides of the original figure by a factor of five. (It's okay if this figure overlaps with others; using a different color might help.)

c)   Reduce the original figure by making its corresponding sides $\frac{1}{2}$ as long.

d)   Make a table with the following headings and complete it for the figures you created in parts (a), (b) and (c).
Recall that $\dfrac{P_{new}}{P_{original}} = \dfrac{\text{Perimeter of new figure}}{\text{Perimeter of original figure}}$.

Figure	Side enlargement or reduction ratio	Perimeter	Area	$\dfrac{P_{new}}{P_{original}}$	$\dfrac{A_{new}}{A_{original}}$
original	1	≈ 13.1	7.5	1	1
a)	3				
b)	5				
c)	$\frac{1}{2}$				
e)	20				
f)	n				

e)   Suppose you were to draw a figure whose sides were enlarged by a ratio of 20 to 1 from the sides of the original. What would its perimeter be? What would its area be? Fill in the table without drawing the figure.

f)   Suppose n to 1 is the side enlargement ratio. Fill in the new perimeter, the new area, and the ratios in the table.

GG-8.   To find the area of an unusually shaped region like the one shown on the right, it often helps to partition the original region into shapes whose areas are easily calculated. Finding the areas of the subregions (or sections) are **subproblems** of the original problem of finding the area of the entire figure.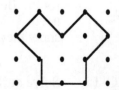

a)   Copy the figure on dot paper and then partition it into rectangles and triangles. (Squares are special rectangles.)

b)   Find the area of each of the subregions you formed. In doing this you are solving subproblems that will contribute to the solution of the original problem.

c)   Now solve the original problem: What is the area of the entire figure?

GG-9.   Solve for y: $4x + 2y = 8$

GG-10.   a)   Graph the quadratic equation $y = x^2 - 5x - 6$ for the values $-2 \le x \le 7$.

          b)   Find all values of x where the graph crosses the x-axis (x-intercepts).

          c)   Solve the equation $x^2 - 5x - 6 = 0$ by factoring.

          d)   What do you notice about your answers to parts (b) and (c)? Write your response in a complete sentence.

GG-11.   Plot the points A (5, 2) and B (0, –1).

          a)   Describe a slide from point B to point A.

          b)   Write the coordinates of another point on the line which contains points A and B.

          c)   Find the length of the segment AB.

GG-12.   Find each of the following products.

          a) $(x - 7)(x - 7)$

          b) $(x + 2)^2$

          c) $3x(5x^2 - 6x - 4)$

          d) $(2x - 5y)^2$

GG-13.   Solve each system of equations for x and y:

          a) $y = 3x - 4$
                $y = \frac{1}{2}x + 7$

          b) $2x + y = 6$
                $y = 12 - 3x$

          c) $y = 0$
                $y = x^2 - 2x + 1$

GG-14.   Your friends are remodeling their house and doubling the length and width of their family room. Because their old wall-to-wall carpeting measured 156 square feet, they think they will need 312 square feet of new carpet. Explain to them, politely, why they are mistaken and show them a short cut for figuring out how much carpet they really need.

GG-15.   Solve each of the following equations.

          a) $\frac{x}{4} = x - 1$ b) $\frac{2x}{3} = x - 4$

          c) $\frac{x}{4} = \frac{x - 2}{3}$ d) $\frac{2x}{3} = \frac{x + 6}{4}$

8.2 ALGEBRAIC PROCESSES AS SUBPROBLEMS

GG-16.   Compare the areas of the shaded regions for the two diagrams below, and write a sentence about the comparison.

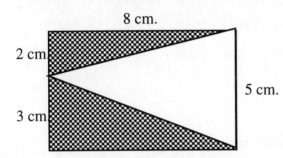   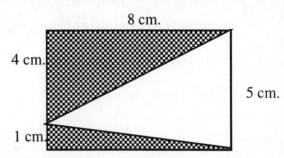

GG-17.   a)   For x = 1, 1.5, and 2, compute the area of the shaded region below.

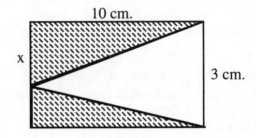

           b)   Find an algebraic expression to represent the shaded area. Then find the area of the shaded region.

           c)   Will your expression work for any length of x? Explain why or why not.

GG-18.   If $3x + 2 = 14$, find $5x + 1$.

           a)   Identify what you should do first. This is the subproblem.

           b)   Find $5x + 1$.

GG-19.   a)   Suppose $3x - 1 = 14$ and $x + y = 7$. Find $2y + 6$. What two subproblems do you need to solve before finding $2y + 6$?

           b)   Solve both subproblems.

           c)   Find $2y + 6$.

GG-20.   Compute each of the following products.

           a) $(\sqrt{36})^2$ b) $(\sqrt{5})^2$

           c) $(\sqrt{x + 2})^2$ d) $(\sqrt{x^2 - 4})^2$

           e) $(x + 3)^2$

GG-21.   Discuss in your group how to rewrite this equation without square roots: $\sqrt{x+1} = 4$.

GG-22.   Tasha was given the equation $\sqrt{x^2 + 6} = x + 2$ and did not know what to do. Her teacher suggested that eliminating the square root symbol was a subproblem.

a)   Tell how you could solve this subproblem.

b)   Solve the subproblem.

c)   Find x.

GG-23.   Tasha was given another problem: Solve the equation $\sqrt{5x+3} + 7 = 10$.

a)   In this case, unlike the previous one, squaring both sides of the equation does not help. The first subproblem is to isolate the square root on one side of the equation. How can Tasha do this?

b)   What would be the next subproblem? Show how to do this.

c)   Find x.

GG-24.   Each of the following algebraic expressions represents the area of one of the figures. Match each algebraic expression with a figure. It is possible that some figures are associated with more than one algebraic expression.

a) $x + 2$

b) $x^2 + 2$

c) $2x^2$

d) x^2

e) 1

f) x

g) $x^2 + x$

h) $3(x + 2)$

i) $x^2 - 1$

j) $x(x + 1)$

k) $x^2 + 2x + 1$

l) $2 - x^2$

m) $3x + 6$

n) $(x + 1)^2$

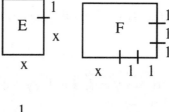

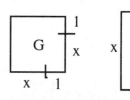

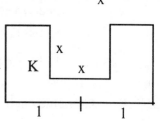

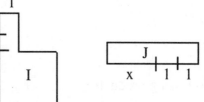

GG-25. a) How many 6" by 6" square tiles will be needed to cover a 12' by 16' rectangular floor?

b) Explain an easy way to see that the answer to part (a) is the same as the answer to GG-3.

GG-26. a) Graph the following linear equations on the same set of axes. Label each graph with its equation. (Before you start, think, "How many points are needed to graph a line?")

$$y = \frac{1}{2}x$$

$$y = \frac{1}{2}x + 2$$

$$y = \frac{1}{2}x - 3$$

b) What is the same about the three graphs? What is the same about the three equations?

GG-27. Rewrite each of the following expressions using as few digits or variables as possible. Assume that $x \neq 0$.

a) $\dfrac{7659}{7659}$

b) $\dfrac{x}{x}$

c) $\dfrac{x + 3}{x + 3}$

d) $\dfrac{12x}{3x}$

e) $\dfrac{16(x + 2)}{4}$

f) $\dfrac{(x + 87)^2}{x + 87}$

g) Does $\dfrac{x + 3}{x} = 3$? Explain why or why not.

GG-28. a) Copy this rectangle and write an algebraic expression for its area two ways:

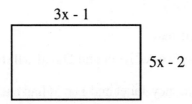

3x - 1

5x - 2

b) For what value of x will the rectangle be a square?

c) If the rectangle is a square, what is its area?

GG-29.   Solve each of the following equations or systems of equations.

   a) $-2(3x + 6) = -8x - 6$ b) $-3x + 7 = 2x + 32$

   c) $6 - 8(3c + 6) = 0$ d) $4x + y = 8$
                                                  $y = 12 - 6x$

GG-30.   Solve for y:

   a) $4x - y = 15$

   b) $4x = 6y - 10$

GG-31.   Factor and solve each of the following equations.

   a) $x^2 - 2x - 48 = 0$

   b) $3x^2 - 6x = 2x^2 + 2x - 15$

   c) $x^2 + 12x = 5x + 8$

GG-32.   Steffi managed to drive from Sacramento to Los Angeles at an average speed of 85 miles per hour. About how long did it take her on the average to drive one mile?

GG-33.   **Thrifty People** Janet has $290 and saves $5 a week. David has $200 and saves $8 a week. In how many weeks will they both have the same amount of money?

   a)   Let x represent the number of weeks Janet saves her money. Write an equation which describes how much money Janet will save, using y to represent her total savings. Label the equation "Janet's savings"

   b)   Use the same variables to write and label a similar equation for David.

   c)   On graph paper, draw and label a set of coordinate axes. Graph and label both savings equations on the axes.

   d)   Use your graph to estimate ...

      (1)   in how many weeks Janet and David will both have the same amount of money.

      (2)   how much money Janet and David had each saved by this time.

   e)   Use your two equations to determine the number of weeks needed for Janet and David to save the same amount of money.

8.3 ALGEBRAIC FRACTIONS AS SUBPROBLEMS

GG-34.   Identify one set of subproblems needed to write 48 as a product of primes. Now solve them!.

GG-35.   Identify one set of subproblems needed to write 5445 as a product of primes, and then solve them.

GG-36.   Write each of the following expressions in simplest fraction form.

a) $\dfrac{5 \cdot 3}{2 \cdot 3}$

b) $\dfrac{5 \cdot 5 \cdot 5}{5 \cdot 5 \cdot 6}$

c) $\dfrac{10}{10^3}$

d) $\dfrac{6x^3}{2x}$

e) $\dfrac{8x^3 y^4}{4x^2 y}$

f) $\dfrac{12(x-2)^2}{3(x-2)}$

g) $\dfrac{6(m+1)^3}{6(m+1)}$

h) Explain in one or two sentences how your group got its solution to (f).

GG-37.   Write each of the following expressions in simplest form.

a) $\dfrac{(x+3)^2}{(x+3)(x-2)}$

b) $\dfrac{8(2x-5)^3}{4(2x-5)^2(x+4)}$

c) $\dfrac{12(x+1)^2 (x-2)^3}{6(x+1)^3 (x-2)^5}$

d) $\dfrac{x^2 + 6x}{x^2 + 12x + 36}$ You must factor first!

GG-38.   a) Copy this rectangle and write an algebraic expression for its area two ways:

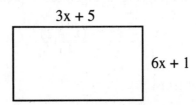

3x + 5

6x + 1

b) Find the area of the rectangle for each of the following values of x: x = 1, x = 2, and x = 3.

GG-39.   Lucia was given the following expression to rewrite in simplest form.

$$\frac{x^2 + 6x}{(x + 6)^2} \cdot \frac{(x^2 + 7x + 6)}{x^2 - 1}$$

After looking at it, she realized it was easy to do because it broke up into five subproblems. The first four subproblems are to factor the four polynomials. Although $(x + 6)^2$ is already factored, it can be rewritten as $(x + 6)(x + 6)$.

She got:
$$\frac{x(x + 6)}{(x + 6)^2} \cdot \frac{(x + 6)(x + 1)}{(x - 1)(x + 1)}$$

Lucia then rewrote it all so that whenever she could, she had the same terms in the numerator and denominator:

$$\frac{(x + 6)}{(x + 6)} \cdot \frac{(x + 6)}{(x + 6)} \cdot \frac{x}{(x - 1)} \cdot \frac{(x + 1)}{(x + 1)}$$

Since $\frac{(x + 6)}{(x + 6)} = 1$ and $\frac{(x + 1)}{(x + 1)} = 1$, the entire messy-looking product became

$$1 \cdot 1 \cdot \frac{x}{(x - 1)} \, 1, \text{ or just } \frac{x}{x - 1}.$$

Thus, $\dfrac{x^2 + 6x}{(x + 6)^2} \cdot \dfrac{(x^2 + 7x + 6)}{x^2 - 1} = \dfrac{x}{x - 1}.$

The important thing Lucia learned was that **each part was easy** and **doing several easy parts allowed her to do a hard-looking problem**.

Can $\dfrac{x}{x - 1}$ be simplified further? Check using one or two values for x. What restrictions are there on the values of x that can be used?

GG-40.   The following problem of simplifying an algebraic expression is best viewed as five subproblems.

$$\frac{x^2 - 3x}{x^2 - 4} \cdot \frac{(x - 2)^2}{x^2 - 9}$$

a)   State the five subproblems.

b)   Do each subproblem you found in part (a).

c)   Simplify the given expression.

GG-41.　Graph the equation $y = \frac{1}{2}x - 2$.

a)　Label the y-intercept.

b)　By completing this sentence, describe the line you graphed in terms of a slide:

"For points on the graph, when the x value increases by 2, the y value ＿＿＿＿＿"

GG-42.　Simplify each of the following expressions by using the same procedure as in GG-40: state the subproblems, do each subproblem, and then use the results to simplify the given expression.

a)　$\dfrac{x^2 + 6x - 7}{x^2 - 3x + 2} \cdot \dfrac{x^2 - 5x + 6}{x + 7}$

b)　$\dfrac{7x + 42}{x^2 + 5x - 6} \cdot \dfrac{x^2 - x}{x^2}$

GG-43.　These two line segments have equal lengths:

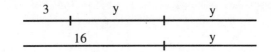

Write an equation to describe the situation, and then solve for y.

GG-44.　Goliath had a 36 mile head start, but David caught up with him in three hours. How fast was David traveling if his speed was twice that of Goliath? Draw a diagram, write an equation, and then solve the equation.

GG-45.　Write a shorter expression for each of the following polynomials.

a)　$(3x^2 + 0.8) - (7x^2 - 5x)$

b)　$(12x^2 - 3x) + (2x^2 - 4x + 3)$

c)　$(7x^2 + 5) - (3x^2 + 2x - 4)$

d)　$(x^2 + 3x + 4) + (3x^2 - 4x - 7)$

e)　$(x^3 + 2x^2 + 4) - (5x^2 - 4)$

GG-46.   Follow Lucia's example in GG-39 and simplify each of the following expressions by factoring and looking for fractions that are equivalent to 1.

a) $\dfrac{x^2 + 2x}{x + 2}$

b) $\dfrac{x^2 + 2x + 1}{x^2 + x}$

GG-47.   Graph the line $y = 4x - 8$.

a)   Find the area of the triangle formed by the line and the x and y axes.

b)   Find the perimeter of the triangle described in part (a).

c)   Write the ratio of the length of the long leg of the triangle to the length of the short leg.

8.4 CIRCLES AS SUBPROBLEMS

The perimeter of a circle is usually called its circumference.
A circle of radius r units has a circumference of length $2\pi r$ units, and it has an area of πr^2 square units.

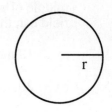

$$\text{circumference } C = 2\pi r \text{ units}$$
$$\text{area } A = \pi r^2 \text{ square units}$$

Check if your calculator has a $\boxed{\pi}$ key to get an approximate decimal value for the irrational number π.

GG-48.   If the area of the whole circle is 100 square centimeters, what is the area of the shaded region?

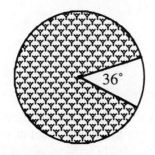

If you need help getting started, answer the following questions:

a)   What must you know in order to do the problem; that is, what are the sub-problems?

b)   How many degrees are there around the center of a circle?

c)   What fraction of the circle is shaded?

GG-49.   What diameter pizza would you have to buy to get at least 100 square inches of pizza? Be sure to draw pictures and identify any subproblems.

GG-50.   For a circle of radius r, find the value of r if ...

a) $308 = 2\pi r$

b) $400 = \pi r^2$

GG-51.   🏠🐕 **The Grazing Goat**   A goat is tied by a rope to
one corner of a barn 15 meters by 25 meters in the
middle of a large grassy field. Over what area of the
field can the goat graze if the rope is ...

a)   10 meters long?

b)   20 meters long?

c)   30 meters long?

d)   40 meters long?

Now suppose that the rope is x meters long.
Express the area over which the goat can graze if ...

e) $0 < x < 15$

f) $15 < x < 25$

g) $25 < x < 40$

GG-52.   A school grass playing field is shaped as shown below. The rounded pieces are each circular
arcs of radius 35 yards. The straight edges meet at right angles. If two pounds of fertilizer
are needed for every 100 square feet of grass, about how many pounds of fertilizer will be
needed for the entire field?

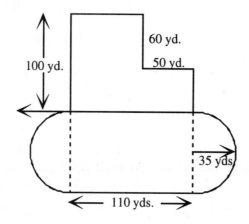

60 yd.

100 yd.

50 yd.

35 yds

110 yds.

GG-53.   a)   Solve for r: $A = \pi r^2$.

b)   Solve for r: $C = 2\pi r$.

c)   For a circle of radius r, what part of the circle is associated with the length 2r ?

GG-54.   Find the area of the shaded region in each of the following figures.

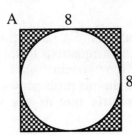

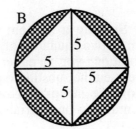

GG-55.   Solve each of the following equations.

a) $\dfrac{3x}{5} = \dfrac{x-2}{4}$ b) $\dfrac{4x-1}{x} = 3x$

c) $\dfrac{4x-1}{x+1} = x-1$ d) $\dfrac{4x-1}{3x+1} = x-1$

GG-56.   The diameter of a flattened Frisbee is 9.2 inches. Find its area and circumference. Be sure to draw pictures and identify any subproblems you use.

GG-57.   The winner of the women's marathon at the 1990 Goodwill Games ran the second half of the race at a pace of six minutes per mile. At what speed, in terms of miles per hour, did she run the end of the race?

GG-58.   Find the area and perimeter of this trapezoid:

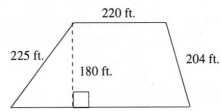

(One subproblem is to find base of a certain right triangle.)

GG-59.   A square cake plate has area 150 square inches. Ida Mae is baking a round cake. What is the maximum area the bottom of a round cake can have and still fit on the plate? Be sure to draw pictures and identify any subproblems.

GG-60.   Solve this problem by any method:
Where do the graphs of $y = x^2 - 4$ and $y = 2x - 1$ intersect?

GG-61.   a)   Solve for x: $\sqrt{x^2 + 5} = 5$.

b)   Solve for x and y: $x + y = 16$ and $2y = x - 4$.

GG-62.   Simplify the expression $\dfrac{x^2(x^2 - 3x + 2)}{x(x-2)}$.

8.5 SOLVING SYSTEMS OF EQUATIONS USING SUBPROBLEMS (ELIMINATION METHOD)

GG-63. **The Elimination Method for Solving a System of Equations** So far in this course, when you've been given a system of two equations of two-variable equations to solve, you've used one of two ways to find a solution: you've either graphed the two equations to find the point(s) of intersection, or you've used the substitution method. In this problem you'll use the strategy of **identifying and solving subproblems** to **eliminate one of the variables to obtain a single equation** containing one variable.

a) Solve this system of equations by graphing:

$$2x + y = 11$$
$$x - y = 4$$

b) Now solve the same system of equations by substitution.

c) There is also another algebraic method that can be used to solve systems of equations -- and it uses subproblems! We know how to solve single-variable equations. When there's a pair of two-variable equations, we can eliminate one of the variables to obtain one single-variable equation. We can do this by adding, as shown below:

To solve the system of equations, we can eliminate the y terms by adding the two equations together,

$$\begin{array}{r} 2x + y = 11 \\ \underline{x - y = 4} \\ 3x + 0y = 15 \end{array}$$

and then solve for x.

$$3x = 15$$
$$x = 5$$

Once we know the x-value we can substitute it into <u>either</u> of the original equations to find the corresponding value of y. Let's use the first equation:

$$2x + y = 11$$
$$2(5) + y = 11$$
$$10 + y = 11$$
$$y = 1$$

We can check our solution by substituting both the x-value and y-value into the other original equation:

$$x - y = 4$$
$$5 - 1 = 4$$
It checks!

What happens if you first substitute the x–value into the second original equation? Try it and check your answer.

Now write the solution to the system as an ordered pair. What does the ordered pair represent in terms of the graph of the two original equations?

d) Which of the three methods for solving the given system of equations --graphing, substitution, or elimination -- was easiest for you to understand and do? Write one or two sentences to explain your response.

GG-64.   Solve the following systems of equations by eliminating one of the variables.

a) $4x + 2y = 14$
     $x - 2y = 1$

b) $5x + 3y = 25$
     $7x - 3y = -1$

c) $3x + y = 13$
     $2x - y = 2$

d) $5x + y = 20$
     $2x + y = 8$

e)   Show two checks to make sure that your answers to part (a) are correct.

GG-65.   What percent of the total area of rectangle ABCD is shaded?

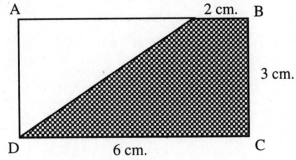

GG-66.   What is the area of the shaded region?

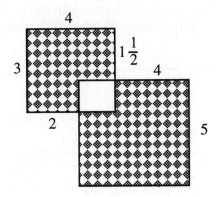

GG-67.   a)   Find the area of the shaded region in terms of r and π.

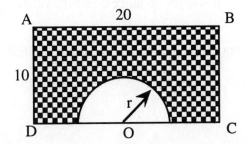

b)   If segment DO is the same length as segment OC, for what values of r is the expression you found in part (a) valid?

GG-68.   Solve each of the following equations by factoring.

   a) $2x^2 + 6x = 8$ b) $x^2 - 7x + 10 = 0$

   c) $3x(x - 5) = 17 - x$ d) $(x - 1)(x - 5) + 4 = 0$

GG-69.   Find each of the following sums of fractions.

   a) $\dfrac{1}{2} + \dfrac{1}{3}$

   b) $\dfrac{1}{2} + \dfrac{x}{3}$

   c) $\dfrac{x}{2} + \dfrac{x}{3}$

   d) $\dfrac{1}{2} + \dfrac{1}{x}$

   e) $\dfrac{1}{3} + \dfrac{1}{2x}$

GG-70.   The diagram below represents the track around a football field. The total distance around the track for Lane 1 is 440 yards. Each straight-away is 100 yards long. The curved ends of the track are semi-circles.

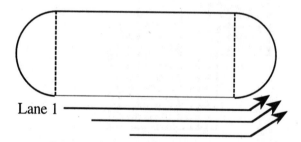

Lane 1

   a)   If each running lane is one yard wide, what distance must a runner in Lane 8 run when going completely around the track?

   b)   How long should the stagger distance be so the runner in Lane 8 runs 440 yards?

GG-71.   In problem GG-71, suppose the area inside the track is entirely covered with grass. Find the area of the grass.

GG-72. Solve each of the following systems of equations. Identify the method you use: graphing, substitution, or elimination.

a) $3x - 2y = 4$
     $4x + 2y = 10$

b) $p + q = 4$
     $-p + q = 7$

c) $y = x + 2$
     $x + y = -4$

GG-73. The length of the diagonal on Ms. Speedi's classroom floor is 30 feet. The height of the room is 10 feet. A fly starts in one corner of the room, corner A, and flies directly to the farthest opposite corner, corner B.

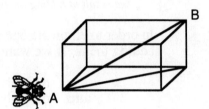

a)   How far does the fly travel?

b)   What is the distance if the fly starts in a different corner and flies to the farthest corner opposite it? Explain.

GG-74. If two 10-inch pizzas cost the same as one 15-inch pizza, which is the better buy? What if the crust is your favorite part of a pizza? Would it make a difference? Explain your answers.

GG-75. Solve each of the following equations.

a) $\frac{x}{3} = x + 4$

b) $\frac{x + 6}{3} = x$

c) $\frac{x + 6}{x} = x$

d) $\frac{4x + 1}{3x + 1} = x + 1$

GG-76. Find x:

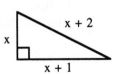

8.6 THE ELIMINATION METHOD USING MULTIPLICATION

GG-77. **The Elimination Method for Solving a System of Equations by Multiplication**
Here's another example of solving a system of equations by first eliminating one of the variables.

Suppose we want to solve this system of equations:

$$3x + 2y = 11$$
$$4x + 3y = 14$$

Again, **the subproblem is to eliminate either x or y** when we add the equations together. However, in this case if we just add the equations, neither x nor y will be eliminated:

$$3x + 2y = 11$$
$$\underline{4x + 3y = 14}$$

the result of adding the two equations is an equation with x and y → $7x + 5y = 25$

In order to eliminate one of the variables we'll need to do something to BOTH equations before adding them. If we want to eliminate y, we could ...

multiply the first equation by 3 to get $9x + 6y = 33$
and   multiply the second equation by –2 to get $-8x - 6y = -28$

Why did we choose to multiply the first equation by 3 ?
Why did we multiply the second equation by –2 ?

What could we do now to eliminate the y terms?
Answer: We can eliminate the y terms by adding the two new equations:

$$9x + 6y = 33$$
$$\underline{-8x - 6y = -28}$$
$$x + 0y = 5$$

Now we know x = 5 and can easily find that y = –2. *How?* The solution is (5,–2).

a) Discuss the example in your group. What would have happened if we had chosen to multiply the first equation by 4 and multiply the second equation by –3 ? Check by writing out all the steps.

b) There are many ways to solve this problem using the idea of multiplying each equation by a number and then adding to get 0x or 0y. Choose another two numbers that will work as multipliers and then write out all the steps. Check your choice of numbers and your work with your group.

GG-78. In the figure on the right, the curves are quarter circles cut out of a rectangle. The radius of the smaller circle is one unit.

a) Find the dimensions of the original rectangle.

b) Find the area of the figure.

c) Find the perimeter of the figure.

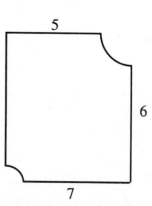

GG-79. Farmer Fran is building a rectangular pig pen alongside his barn. He has 100 feet of fencing, and he wants the largest possible area in which the pigs can muck around. What should the dimensions of the pig pen be?

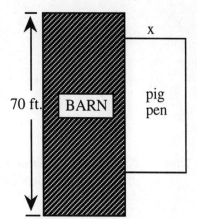

a) Use a Guess and Check table to compare the width and area of the variuos pens.

b) Write an expression for the length of the pen if x is the width.

c) Write an expression for the area of the pen if x is the width.

d) Use the values in your Guess and Check table and some additional values to draw a graph of the area in terms of x.

e) Use your graph to solve Farmer Fran's problem.

GG-80. a) On graph paper draw four different rectangles, each with a perimeter of 30 units.

b) Make a table of all rectangles which have perimeter 30 units and integer lengths for the base and altitude. Include the area of each rectangle in your table.

c) Graph your data from part (b) with "altitude" along the horizontal axis and "area" along the vertical axis.

d) Use your graph from part (c) to estimate the maximum area for a rectangle with perimeter 30 units.

e) What was the maximum perimeter of the rectangles?

GG-81. Graph the equation $y = \frac{1}{2}x^2 - 6$ for $-5 \le x \le 5$.

GG-82. Use the elimination method to solve each of the following systems of equations. In each case you will have the subproblem of eliminating either x or y from the equations, and will need to do something before adding. What could you do? There is more than one way to solve these, so expect a <u>variety of methods</u> in your group, but the <u>same solutions</u>, (x, y).

a) $2x + 3y = -1$
    $5x - 2y = -12$

b) $2x + 3y = 17$
    $x + 3y = 16$

c) $2x + y = 7$
    $2x + 5y = 12$

d) $x + 3y = 4$
    $3x - y = 2$

e) Write out the two checks that your solution to part (b) is correct.

GG-83.   Solve each system of equations.for x and y.

    a) $x + 2y = 5$
           $x + y = 5$

    b) $2x + 3y = 5$
           $x + 3y = 4$

    c) $x + 2y = 16$
           $x - y = 2$

    d) $3x + 2y = 11$
           $4x - 2y = 2$

    e) $3x + 3y = 15$
           $x - y = 6$

    b) $2x^2 - y = 5$
           $x + 2y = -5$

GG-84.   **Chapter 8 Summary: Rough Draft** The four main ideas of the chapter were mentioned in the chapter Introduction. Look back through the chapter to find where each main idea was developed.

Write your answers to the following questions in rough draft form on separate sheets of paper and be ready to discuss them with your group at the next class meeting. Focus on the content, not neatness or appearance, as you write your summary draft. You will have the chance to revise your work after discussing the rough draft with your group.

a)   In this chapter you've worked on many problems where you identified and solved subproblems to solve a larger problem. Select a problem from each section that you think illustrtes this idea well. Write complete sentences to describe how you did each of the selected problems. Then tell why you chose the problems that you did.

b)   Select a problem you didn't understand before, but now know how to do. Show all your work and a complete solution. Explain why you chose the problem.

c)   Select a problem that you liked best or most enjoyed solving. Write the problem and your complete solution. Explain why you chose the problem.

8.7 SUMMARY AND REVIEW

GG-85. **Chapter 8 Summary: Group Discussion** Take out the rough draft summary you completed in GG-84. Use this time to discuss your work and use homework time to revise your summary as needed.

Take turns to describe the problems group members chose to illustrate their understanding of how subproblems are used to solve larger problems. Each group member should:

- explain the problems he or she chose to illustrate this main idea;

- explain why he or she chose those problems; and

- explain a problem he or she didn't understand before, but now can solve, and explain why the problem was chosen.

This is your chance to make sure your summary is complete. You could also update your Tool Kits, if needed, and work together on problems you may not have been able to solve.

GG-86. Solve for x: $\sqrt{3x - 1} = 18$

GG-87. Simplify: $\dfrac{3x^2 + 6x + 3}{x^2 + 3x + 2}$

GG-88. Solve each of the following systems of equations.

a) $2x - y = 16$
    $x + y = 14$

b) $x^2 - y = 9$
    $3x + y = 19$

c) $2x - y = 16$
    $3x + 4y = 24$

GG-89. Find the area of a right triangle with a hypotenuse of length 13 centimeters and one leg of length 5 centimeters. What subproblem did you need to solve?

GG-90. A rectangle is three times as long as it is wide. If the length and width are each decreased by four units, the area is decreased by 176 square units. What are the dimensions of the original rectangle?

GG-91.   A square with sides of length x centimeters has a circle cut out
         of it. The circle has a diameter x centimeters long.   Find the
         fraction of the square that is left if ...

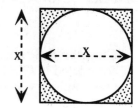

a)   x = 2

b)   x = 20

c)   x = 10

d)   you just know that the diameter is x (Write the fraction in terms of x.)

GG-92.   Suppose a rectangle has one side of length 7 centimeters and a diagonal 10 centimeters long.
         What is the area of the rectangle? In order to solve this problem, you need to find the length of
         the other side of the rectangle. This is a subproblem.

a)   Draw a diagram and use the picture to help you find the length of the other side of the
       rectangle.

b)   Find the area of the rectangle.

GG-93.   For Katja's party, Lawson, Andrea, Samantha, and Ryan are going to hang a piñata half way
         between two poles with a rope over the top of each pole. Both poles are 20 feet high, and
         they are 15 feet apart. The piñata must hang four feet above the ground. How much rope is
         needed to hang the piñata? Draw a diagram before you start.

GG-94.   **Chapter 8 Summary: Revision** Use the ideas your group discussed in GG-85 to revise
         your Chapter 8 Summary. Your presentation should be thorough and organized, and should
         be done on paper separate from your other work.

GG-95.   **Course Summary Update** You've learned a lot in this course so far and it is useful to pull
         it all together. Take some time now to revise the Course Summary you wrote at the end of
         Chapter 6 to include important ideas you've learned in the last two chapters. Include copies of
         homework problems that illustrate these important ideas.

Algebra Tool Kit: the "what to do when you don't remember what to do" kit

Algebra Tool Kit: the "what to do when you don't remember what to do" kit

Dot Paper

Dot Paper

Dot Paper

Chapter 9

THE BURNING CANDLE: More Ratios and Slope

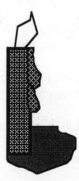

CHAPTER 9

THE BURNING CANDLE:
MORE RATIOS AND SLOPE

THE
BURNING
CANDLE

Suppose it's your friend's birthday and you want to surprise her by walking into the room carrying a piece of cake with a lighted candle. Could you predict how long before the candle goes out? (You may have worked on a problem similar to this one in the Burning Candle Investigation from Chapter 2, where you estimated how long it would take for the candle to extinguish by graphing data points.)

To answer the question, you could collect some data: let a candle burn for 50 seconds and weigh it. Then let it burn for 3 more minutes (for a total of 3 minutes 50 seconds). Suppose the candle weighs 0.78 grams after 50 seconds, and weighs only 0.57 grams after 3 minutes 50 seconds. If this is the only information you have, can you predict when the candle will go out? What assumptions would you make?

In this chapter you will be using the concept of **ratio** to learn techniques for writing equations from data or graphs. First, you'll review the concept of ratio and then look at some applications of ratios.

In this chapter you will have the opportunity to:

- review ratios;

- develop a good understanding of what the slope of a line represents and be able to estimate slopes of arbitrary lines;

- understand that slope is a ratio and be able to relate computing slopes to similar right triangles;

- be able to graph using the slope-intercept form of a line; and

- be able to write the equation of a line given either a point and a slope, or two points.

CHAPTER CONTENTS

9.1 RATIOS AND THE SLOPE OF A LINE

BC-1.   A line contains the point (3, –2) and the origin. Plot the two points on graph paper, but **do not draw in the line**. Use your picture to find four other points with integer coordinates that are on the line.

BC-2.   The two points (3, 1) and (4, 3) lie on a line. Describe a slide you could use to find five other points with integer coordinates on the line.

BC-3.   Imagine that a classmate who was absent today has called you for help. Given a line through two points -- say (0, 1) and (–2, –3) -- she'd like to know how to find other points on the line that have integer coordinates. Oh, and she wants to be able to find these other points without drawing the line. Use what your group learned in solving BC-1 and BC-2 to explain to your classmate a method for using two points on a line for determining more points on the line.

BC-4.   **y-intercept and x-intercept**   In Chapter 6 you were introduced to the ideas of the y-intercept of a line and the x-intercepts of a parabola. We can apply these the ideas to any graph.

Definition: The **x-intercept** of a graph is the point where the graph crosses the x-axis; that is, where y = 0.
Similarly, the **y-intercept** of a graph is the point where the graph crosses the y-axis, which happens when x = 0.

a)   Add the definitions for x-intercept and y-intercept of a graph to your Tool Kit.

The equation $2x - 3y = 7$ is the equation of a line.

b)   Find the coordinates of the line's x-intercept.

c)   Find the coordinates of the line's y-intercept.

d)   If a point on this line has an x-coordinate of 10, what is its y-coordinate?

BC-5.   The graph of the equation $x - 2y = 4$ is a line.

a)   Find the coordinates of the x- and y-intercepts.

b)   Use the x- and y-intercepts to graph the line.

BC-6.   Read each graph from left to right to compare the graphs of lines *l, m, n* and *t*.

   a)   Use complete sentences to describe how the graphs are similar.

   b)   Use complete sentences to describe how the graphs are different.

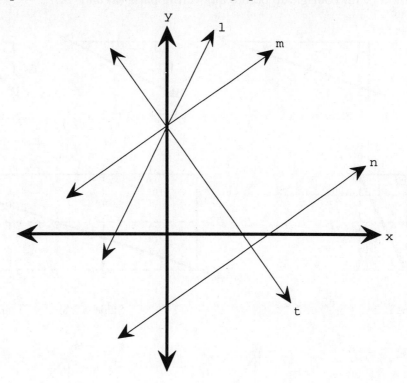

BC-7.   Some of the lines below are quite steep; others are not. Below each line there is a ratio that describes how steep it is. This ratio is called the **slope of the line**.

Compare the following graphs to discover how to determine the slope of a line. Discuss your methods with your group before answering parts (a) and (b).

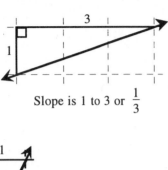

Slope is 1 to 3 or $\frac{1}{3}$

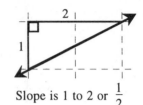

Slope is 1 to 2 or $\frac{1}{2}$

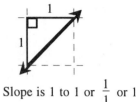

Slope is 1 to 1 or $\frac{1}{1}$ or 1

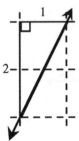
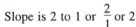

Slope is 2 to 1 or $\frac{2}{1}$ or 2

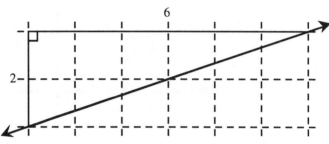

Slope is 2 to 6 or $\frac{2}{6}$ or $\frac{1}{3}$

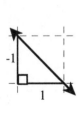
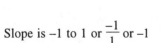

Slope is –1 to 1 or $\frac{-1}{1}$ or –1

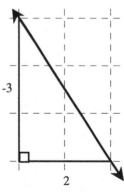

Slope is –3 to 2 or $\frac{-3}{2}$

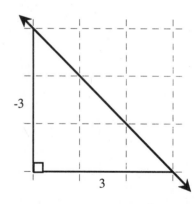

Slope is –3 to 3 or $\frac{-3}{3}$ or –1

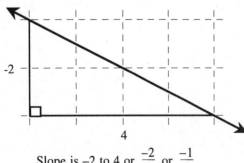

Slope is –2 to 4 or $\frac{-2}{4}$ or $\frac{-1}{2}$

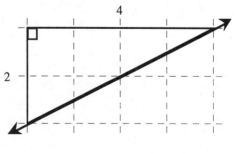

Slope is 2 to 4 or $\frac{2}{4}$ or $\frac{1}{2}$

[PROBLEM CONTINUED ON NEXT PAGE]

BC-7. continued

a) Describe how to find the slope of a line. If your group found more than one method, describe each method. Your descriptions should be clear enough for other groups to understand.

b) Does it matter where the slope triangle is drawn, above the line or below it? Explain why or why not.

c) Use a transparent graph paper grid (or a piece of graph paper) to estimate the slope of each of the lines below. Remember to describe each slope with a positive or negative ratio.

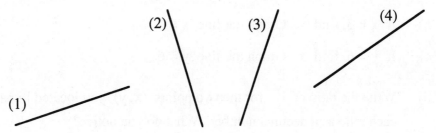

BC-8. **The Slope of a Line** Read the following definition and then copy it into your Tool Kit:

Definition: The **slope** of a line is defined to be

$$\textbf{slope} = \frac{\text{vertical change}}{\text{horizontal change}} = \frac{\text{change in } y \text{ values}}{\text{change in } x \text{ values}}.$$

Note that lines going **upward** ↗ from left to right have **positive** slope, while lines going **downward** ↘ from left to right have **negative** slope. The slope of a line is often denoted by the letter **m**.

To calculate the slope of a line, pick two points on the line and make the segment between them the hypotenuse of a right triangle. For example, if the two chosen points are (–2, 3) and (3, 5), then

$$\textbf{m} = \textbf{slope} = \frac{\text{change in } y \text{ values}}{\text{change in } x \text{ values}} = \frac{2}{5}.$$

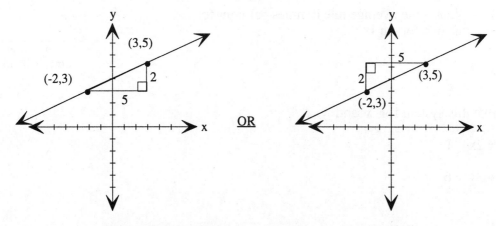

[PROBLEM CONTINUED ON NEXT PAGE]

BC-8. continued

Plot each of the following pairs of points on graph paper, and then determine the slope of the line passing through each pair by counting units.

a)   (1, 2) and (4, 5)   d)   (7, 3) and (5, 4)

b)   (3, 8) and (5, 4)   e)   (–4, 5) and (5, 4)

c)   (–6, 8) and (–4, 5)   f)   (5, 0) and (0, 1)

BC-9.   a)   Graph the line $y = 1.6x$.

b)   If $x = 3$, find y. Graph the line $x = 3$.

c)   If $y = 6$, find x. Graph the line $y = 6$.

d)   Write the ratio of $\frac{y}{x}$ for the two points (x, y) you located in parts (b) and (c). Write each ratio as a decimal number. What do you notice?

BC-10.   Ginger got "1" for the slope of the line through points $(1, 2)$ and $(4, -1)$. Explain to Ginger the mistake she made and how to correctly find the slope.

BC-11.   Where does the line $y = 2x + 1$ intersect the curve $y = x^2 - 2$?

a)   Estimate the point(s) of intersection by graphing.

b)   Solve the problem algebraically.

BC-12.   Use the graph at the right to answer the following questions.

a)   Which car is traveling at the greater rate?

b)   What is the average rate in miles per hour for car A ? for car B ?

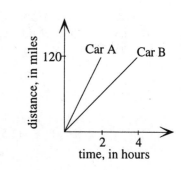

BC-13.   Solve this system for x and y:

$y = 2x - 1$

$x + 3y = 6$

BC-14.   The Oh-So-Good Café makes 15% profit on its lunches and 22% profit on its dinners. If the Oh-So-Good took in $2,700 on Tuesday and made $524 profit, how much was spent at lunch? Let x represent the sales at lunch and y represent the sales at dinner, write two equations, then solve.

BC-15.   Solve each of the following equations for x.

a) $\dfrac{x}{6} = \dfrac{7}{3}$

b) $\dfrac{6}{x} = \dfrac{7}{3}$

c) $\dfrac{6}{x} = \dfrac{4}{x+1}$

d) $\dfrac{x}{x+1} = \dfrac{7}{3}$

BC-16.   In each part below, find the length of side BC, if $\triangle ABC$ is similar to $\triangle DEF$.

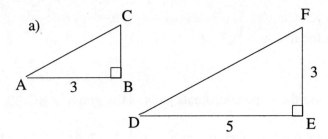

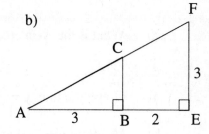

c)   For the drawing in part (b), if the coordinates of point A were (0, 0), what would be the coordinates of points C and F ?

BC-17.   Find each of the following sums.

a) $\dfrac{2}{3} + \dfrac{1}{5}$

b) $\dfrac{1}{x} + \dfrac{1}{3}$

c) $\dfrac{2}{x} + \dfrac{3}{5}$

d) $\dfrac{3}{x} + \dfrac{5}{2x}$

e)   Solve the equation: $\dfrac{3}{x} + \dfrac{5}{2x} = \dfrac{1}{6}$

9.2   SLOPES OF PARALLEL LINES AND THE SLOPE-INTERCEPT PATTERN IN   Y = MX + B

BC-18.   Use dot paper for this exercise. Label a point A in the middle of the paper. Use a straight edge to carefully draw lines through point A with each of the following slopes:

$$1, 2, 3, 4, 5, -1, -2, -3, -4, \frac{1}{2}, \frac{3}{5}, \frac{1}{10}, \frac{-1}{8}, \frac{-2}{3}, \text{ and } 0.$$

Write each line's slope next to it.

BC-19.   **Slopes of Segments**   Use your copy of the resource page for BC-19 to draw each of the following segments: DE, ED, DF, GH, CF, EF, and EG. Next to each segment, write its slope.

BC-20.   a)   Graph the line $y = \frac{1}{2}x$.

          b)   Mentally verify that this line goes through the points $(0, 0)$, $(2, 1)$, $(4, 2)$, $(10, 5)$, $(-6, -3)$, and $(-8, -4)$. Explain how you could do this for point $(2000, 1000)$.

          c)   Compute the slope of the line $y = \frac{1}{2}x$.

          d)   What is the y-intercept of the line $y = \frac{1}{2}x$?

BC-21.   a)   Graph the line $y = \frac{1}{2}x + 3$ on the same coordinate axes as the graph in BC-20.

          b)   Mentally verify that this line goes through the points $(0, 3)$, $(2, 4)$, $(4, 5)$, $(10, 8)$, $(-6, 0)$, $(-8, -1)$, and $(2000, 1003)$.

          c)   Compute the slope of the line $y = \frac{1}{2}x + 3$.

          d)   What is the y-intercept of the line $y = \frac{1}{2}x + 3$?

          e)   Do the two lines represented by $y = \frac{1}{2}x$ and $y = \frac{1}{2}x + 3$ ever intersect?

          f)   Write an equation for another line that is parallel to these two lines.

          g)   Do these lines slope upward or downward?

BC-22.   a)   Find two points on the graph of $y = -\frac{3}{2}x + 4$.

         b)   Find the slope of this line using the two points you found on the graph in part (a).

         c)   Does the line slope upward or downward?

         d)   Where does the graph cross the y-axis?

         e)   Write an equation for a line that is parallel to this line.

BC-23   **Slope and y-intercept in $y = mx + b$** Look again at the equations of the lines in problems BC-20, BC-21 and BC-22. Each equation is written in the y-form $y = mx + b$, where m and b are numbers.

For example, in the linear equation $y = \frac{1}{2}x + 3$, the coefficient of x is $\frac{1}{2}$, so $m = \frac{1}{2}$, and the constant term is 3, so $b = 3$.

         a)   In your group, state as clearly as you can a rule for a fast and sure way of knowing the slope of a line in the form $y = mx + b$.

         b)   Now state a rule for finding the y-intercept of a line in the form $y = mx + b$.

         c)   Add your rules for finding the slope and the y-intercept of a line in the form $y = mx + b$ to you Tool Kit.

BC-24.   Use your rules from BC-23 to state the slope and y-intercept of each of the following lines.

         a) $y = 2x - 5$

         b) $y = -3x + 10$

         c) $y = \frac{3}{4}x + \frac{7}{4}$

         d) $2y = 4x - 5$ (Hint: First solve for y.)

         e) $4y = 3x + 6$

         f) $y + 2x = 4$

         g)   Two **pairs** of lines above are parallel. Write their equations.

BC-25.   A line has the equation $y = \frac{2}{3}x - 1$.

a)   Verify that the points A (3, 1), B (6, 3) and C (15, 9) are all on the line.

b)   Compute the slope of the line using the points A and B.

c)   Compute the slope of the line using the points A and C.

d)   Compute the slope of the line using the points B and C.

e)   What do the results of parts (b), (c) and (d) tell you about using different points to calculate the slope of a line?

BC-26.   A line contains the points (–3, 2) and (2, 5). Elaine thinks that the point (12, 12) is also on this line. Do you agree? If so, tell why. If not, tell how you know it's not on the line. (Hint: Find other points that lie on the line.)

BC-27.   Solve each of the following equations for x.

a) $x^2 - 4 = 3$

b) $(x - 2)(x - 3) = 0$

c) $\frac{x + 2}{3} = \frac{5x + 6}{7}$

d) $\frac{8}{x} + 6 = 2x$

e) $\sqrt{x + 6} - 2 = 6$

f) $x^2 - 2x - 48 = 0$

BC-28.   The length of a certain rectangle is five times its width. Use this information to sketch and label the rectangle. Write an expression to represent the perimeter. Find the rectangle's dimensions if its perimeter is 36 centimeters.

BC-29.   In 35 minutes, Suki's car goes 25 miles.

a)   If she continues at the same speed, how long will it take Suki to drive 90 miles?

b)   Determine the car's average speed in miles per hour.

BC-30.   You solved the following problem in Chapter 3 by making Guess and Check table to write an equation in one variable and then solving the equation. Now you know how to solve equations involving two variables. Draw a diagram and use it write two equations that describe the situation. You could let S represent the length of a short piece and let L represent the length of a long piece. Then solve the problem by solving your system of equations.

A stick 152 centimeters long is cut up into six pieces: four short pieces, all the same length, and two longer pieces, both the same length. A long piece is 10 centimeters longer than a short piece. Into what length pieces is the stick cut?

BC-31.   For the equation $2y - 4x = 6$ ...

         a)   write the coordinates of the x- and y-intercepts.

         b)   graph the intercepts and draw the line.

         c)   find the slope of the line.

         d)   find the distance between the x- and y-intercepts.

BC-32.   Two passenger trains started toward each other at the same time from towns 288 miles apart and met in three hours. The rate of one train was six miles per hour slower than that of the other. Find the rate of each train.

BC-33.   Simplify each of the following expressions.

         a) $(x^2)^3 \cdot (3x^4)^2$

         b) $\dfrac{4x^3y}{12xy^3}$

         c) $\dfrac{2x^2 - x}{x} \cdot \dfrac{1}{2}$

BC-34.   Solve each of the following systems for x and y.

         a) $2x + y = 7$
                 $3x - 2y = 7$

         b) $y = 0.5x + 4$
                 $2x - 4y = 1$

         c)   If you graphed the lines from part (b), where would they intersect?

9.3 PRACTICE WITH Y = MX + B

BC-35.   **This exercise is to be done with a graphing calculator.** You will also need your
copy of the resource page for BC-35.

Interesting patterns often occur when graphing several equations on one set of coordinate axes.
Try these examples, then create some combinations on your own.

In your graphing calculator, enter each set of equations on the same set of coordinate axes. On
your resource page, **sketch** the graphs on one set of axes so you will have a record of them.
Be sure to label line with its equation. Look for differences and similarities among the graphs.

a) $y = x$
$y = 2x$
$y = 4.9x$

What happens to the graph of $y = x$ when the coefficient of x is greater than 1 ? that is,
when x is multiplied by a number greater than 1 ?

b) $y = x$
$y = \frac{1}{2}x$

$y = \frac{1}{4}x$

$y = \frac{1}{6}x$

What happens to the graph of $y = x$ when the coefficient of x is between 0 and 1 ?

c) $y = x$
$y = \frac{-1}{4}x$

$y = \frac{-1}{2}x$

$y = -4x$
$y = -2x$

What happens to the graph of $y = x$ when the coefficient of x is negative?

d) $y = x$
$y = x + 2$
$y = x - 2$
$y = x + 4$
$y = x - 4$

e)   How are all of the equations in parts (a) through (d) alike?

f)   What happens to the graph of $y = x$ when a constant (number) is added to x ?

[PROBLEM CONTINUED ON NEXT PAGE]

BC-35. continued

g)   Identify the slope and the y-intercept for each of the following equations.

y = x	slope = ___	y-intercept = ___
y = 0.5x	slope = ___	y-intercept = ___
y = 0.5x + 5	slope = ___	y-intercept = ___
y = –0.5x	slope = ___	y-intercept = ___
y = 0.5x – 5	slope = ___	y-intercept = ___
y = –0.5x – 5	slope = ___	y-intercept = ___

BC-36.   Interesting patterns often occur if you graph several equations on one set of coordinate axes. The goal here is to find how each equation is related to its graph.

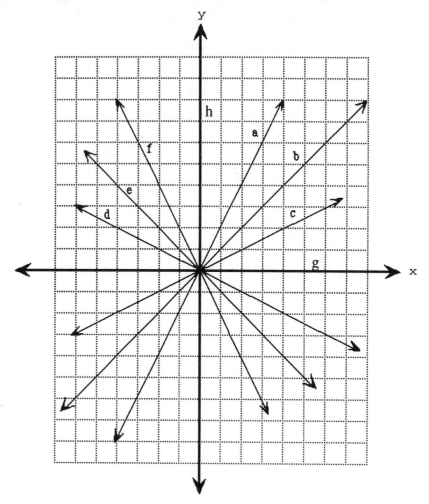

a)   Find the slope of each of the lines (a) through (g) above. Below each line, write its equation in "y = mx + b" form.

b)   Nancy is a new student in Ms. Speedi's class. Explain to Nancy in writing what happens to the graph of y = x when x is multiplied by a positive number greater than 1.

c)   Yasica has just returned to class after a week-long absence. Explain to her, in writing, what happens to the graph of y = x when x is multiplied by a negative number less than −1.

d)   On the coordinate grid above, there are eight lines: the two axes and lines (a) through (f). Between which two of the eight lines would the graph of y = 100x lie? Justify your answer.

e)   Explain to Bernie, another new student, what happens to the graph of y = x when x is multiplied by a number between 0 and 1.

[PROBLEM CONTINUED ON NEXT PAGE]

BC-36. continued

    f)   Continue the explanation to Bernie with what happens to the graph of $y = x$ when x is multiplied by a number between 0 and -1.

    g)   Between which two of the lines would the graph of $y = \frac{2}{5}x$ lie? Justify your answer. Remember, one of the lines may be one of the axes.

    h)   Between which two of the lines would the graph of $y = \frac{3}{2}x$ lie? Justify your answer.

BC-37. a) In the figure below, estimate the y-intercept of each line.

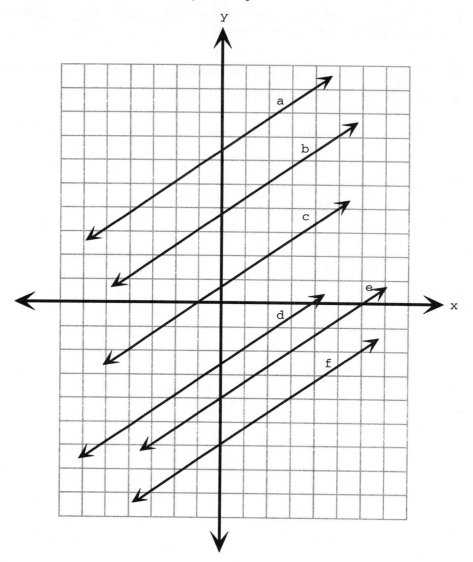

b) What do lines a through f have in common?

c) Between which two lines would the graph of $y = \frac{2}{3}x + 2$ lie?

d) Between which two lines would the graph of $y = \frac{2}{3}x - \frac{12}{7}$ lie?

e) Explain what happens to the graph of $y = \frac{2}{3}x$ when a positive number is added to the right hand side of the equation. What about when a negative number is added to the right hand side of the equation?

f) Write an equation for each of lines a through f.

BC-38.   Is $\dfrac{x-2}{x+3} = \dfrac{-2}{3}$? Explain why or why not.

BC-39.   Solve each of the following quadratic equations.

     a) $x^2 - 7x + 10 = 0$

     b) $9y^2 = 49$

     c) $-x^2 - 11x = 30$

BC-40.   Complete each expression in <u>three</u> different ways so that the resulting expression will factor. Then write each of your three expressions as a product of its factors.

     a) $x^2 + 8x + \underline{}$

     b) $x^2 + \underline{} \cdot x - 10$

BC-41.   Solve each of the following systems of equations for x and y.

     a) $0.7x - 0.3y = 14$
           $2x + y = 9$

     b) $y = x^2 - 7$
           $y = 8 + 2x$

BC-42.   Reggie bragged that he saved $12.60 by buying his new sneakers on sale at 30% off. How much did the sneakers cost originally?

BC-43.   Solve each of the following equations. You may have more than one correct value of x.

     a) $4x + 3 = 1$

     b) $\dfrac{4}{x} + 3 = 1$

     c) $4x^2 + 3x = 1$

     d) $(x+4)(x+3) = 0$

     e) $(x+4)(x+3) = 2$ Before you start: Does the zero product property apply to this equation?

     f) $\sqrt{4x+3} = 1$

BC-44.   Use algebraic methods to convert each of the following equations to **y = mx + b** form.

a) $2y = 4x + 8$

b) $5x - y = 8$

c) $y - 8x = 7$

d) $x = \dfrac{2y-1}{3}$

e) $6x + 3y - 5 = 0$

f) $\dfrac{2}{y} = \dfrac{6}{x+3}$

g)   Use the y-intercept and the slope to graph the equations in parts (a), (e) and (f) <u>without</u> making a table of values.

BC-45.   The equation of a line can be written in slope-intercept form, $y = mx + b$.

a)   For what value(s) of m does the graph slope upward?

b)   For what value(s) of m is the line horizontal?

c)   For what value(s) of m does the line slope downward?

BC-46.   In this problem, write two equations to describe the situation, and solve.

Andy has a 96 inch long piece of wood. From it he needs to cut two pieces of wood as long as possible, and one of the pieces must be 13 inches longer than the other. How long will each piece be?

BC-47.   A certain rectangle has a diagonal of length 15 centimeters, and the ratio width:length:diagonal is 3:4:5 .

a)   Find the width and length of the rectangle.

b)   Find the area of the rectangle.

BC-48.   Arturo has earned 134 points so far in his algebra class. To get an A he needs to earn 90% of the total possible points. One more test, worth 45 points, is coming up. It will bring the total number of points to 195. How many points must Arturo get on this test in order to earn an A in the class? Write an equation and solve the problem.

9.4 USING THE GRAPH TO WRITE AN EQUATION OF A LINE

BC-49. **Are Slope and Segment Length Related?** Use the dot paper shown below to explore the relationship between the slope of a line segment and the length of the segment.

a) Find the slopes of segments AB and CD.

b) Use the Pythagorean theorem to find the lengths of segments AB and CD.

c) Does the length of a line segment seem to be related to the slope of the segment? Explain your answer.

BC-50. The points $(2, 334)$ and $(788, 934)$ lie on a line. Find the coordinates of a third point on the line. How can you use the idea of slope to help you do this?

BC-51. Each pair of ordered pairs below represents two points on a line. In each case …

i) Use graph paper to plot the pair of points on one set of axes and carefully use a straightedge to draw the line through them.

ii) Estimate where each line crosses the y-axis to get the y-intercept.

iii) Compute the slope of each line.

iv) Write an equation of each line in the $y = mx + b$ form.

a) $(3, 4)$ and $(4, 6)$

b) $(1, -5)$ and $(-1, 1)$

c) $(-5, -8)$ and the origin

d) $(1, -5)$ and $(-2, 0)$

BC-52. Draw a sketch of two line segments which have the same slope but are not part of the same line.

BC-53. Is the following situation possible? Justify your response.

Segments UC, CD, and UD have slopes $\frac{1}{2}$, $\frac{1}{2}$, and -2, respectively, **and** points U, C, and D all lie on the same line.

BC-54.   Each of the following equations is in $y = mx + b$ form. For each equation, state the value of m and the value of b.

   a) $y = -4x + 7$

   b) $y = 4x$

   c) $y = 4$

   d) $y = x - \dfrac{3}{2}$

   e)   Complete this statement and add it to your Tool Kit:

        In an equation of the form $y = mx + b$, the m represents the _____ of the line and the b represents _____.
        This is why the form $y = mx+b$ is called the **slope-intercept form** of a linear equation.

BC-55.   Points $(-23, 345)$ and $(127, 311)$ are two points on a line. Find the coordinates of a third point on the line.

BC-56.   a)   How **long** is the line segment from the point $(4, 3)$ to the point $(-2, 7)$? First write your answer in simple square root form, and then as a decimal.

   b)   What is the **slope** of the line segment?

   c)   Does the length of the segment equal the slope of the segment?

BC-57.   Graph each of the following lines on the same set of coordinate axes. What do you notice?

   a) $y = 2x + 3$

   b) $2x - y + 3 = 0$

   c) $4x - 2y + 6 = 0$

BC-58.   Seymour has $2000. He put part of it in an account yielding 6% annually. He put the rest in a restricted savings account yielding 8% annually. After one year the two accounts will have a total of $2136.50. How much money did Seymour invest at 6% ?

BC-59.   Find the area of a rectangle with one side 10,000 millimeters long and the other side 20 meters long. (Hint: think carefully about the units.)

BC-60.   In this problem, write two equations to describe the situation, and solve.

At the Wheatland City College Annual Halloween Costume Contest, the judges and their guests were given 24 free tickets, and 510 tickets were sold. Student tickets were $2 each and general public tickets sold for $3 each. The college took in $80 more on general public tickets than they did on student tickets. How many student tickets were sold?

BC-61.   Solve each of the following equations for x.

a) $1234x + 23456 = 987654$

b) $\frac{10}{x} + \frac{20}{x} = 5$

c) $x(x - 1)(x - 2) = 0$

d) $5x^2 - 6x + 1 = 0$

BC-62.   Use the figure on the right to answer the following questions.

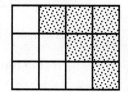

a)   What is the area of the <u>unshaded</u> region? (We'll call the unshaded region a "three-step" region.)

b)   Draw a five-step region and find its area.

c)   What would the area of a ten-step region be?

d)   What would the area of a 157-step region be?

e)   Describe how you would find the area of a region with n steps.

BC-63.   Triangle RAT is similar to △FNK with ∠R = ∠F, ∠A = ∠N, RA = 10 centimeters, RT = 20 centimeters, and AT = 18 centimeters. If FN = 15 centimeters, find NK.

BC-64.   The incidence of rabies in skunks was recently reported to be three of every 10 skunks in a mountain county. Sampling studies (similar to fish sampling) revealed that there are about 22,400 skunks in the county. About how many carry rabies?

9.5 THE BURNING CANDLE: WRITING AN EQUATION OF A LINE GIVEN TWO POINTS

BC-65.   Identify the slope and the y-intercept of each of the following lines by rewriting each equation in **y = mx + b** form.

a) $3x + 2y = 7$

b) $4x - 2 = 3y$

c) $\dfrac{3}{x} = \dfrac{5}{y}$

BC-66.   **How to find an equation of a line without graphing** In BC-51 you found an equation of the line passing through two points by drawing a graph, using the slope triangle, and estimating where the graph crossed the y-axis. But what if the y-intercept were 396 or $1\frac{3}{7}$?

In such cases it would be hard to use a graph to find an equation, so that is when to use the algebra you know. To develop an algebraic method of finding an equation of the line passing through two points, we'll examine a simple situation.

Example: Find the equation of the line passing through the points (3, 4) and (4, 6).

First Step
We know we can write an equation of a line like this:

$$y = mx + b$$
where m is the slope,
and b is the y-intercept.

Second Step
We already know how to calculate the slope by drawing a generic slope triangle:

$$m = \frac{2}{1} = 2$$

So the equation must look like $y = 2x + b$.

$$y = 2x + b$$

Third Step
The hard part is to find the y-intercept, b, without drawing a graph. But remember, we know that (3, 4) is on the line. This means that x = 3 and y = 4 fit in the equation:

$$y = 2(x) + b$$
$$4 = 2(3) + b$$
$$4 = 6 + b$$
$$-2 = b$$

Final Step
Using the values we found for the slope, m, and the y-intercept, b, we can write the equation:

$$y = 2x - 2.$$

In the third step we used (x, y) = (3, 4). See what happens if you use the point (4, 6) instead in the third step of the example. Explain your results in a sentence.

Also, check your equation by drawing a line through the two points (3, 4) and (4, 6) and marking the y-intercept. Do the slope and y-intercept numbers in the equation actually fit the line?

BC-67. Use the method described in BC-66 above to find an equation of the line passing through each pair of points. In each case write the equation in the slope-intercept $(y = mx + b)$ form.

a) $(4, 6)$ and $(6, 7)$

b) $(9, 8)$ and $(3, 5)$

c) $(-3, 2)$ and $(4, 5)$

d) $(5, 342)$ and $(30, 2047)$

BC-68. a) A certain line with slope $\frac{1}{2}$ goes through the point $(6, 1)$. Find an equation of the line.

b) A certain line with slope $\frac{2}{3}$ goes through the point $(-3, 5)$. Find an equation of the line.

c) Suppose the line $y = 2x + b$ goes through the point $(1, 4)$. Find an equation of the line.

d) Suppose the line $y = \frac{1}{2}x + b$ goes through the point $(2, 3)$. Find an equation of the line.

BC-69. Find an equation of each line described below.

a) the line through points $(-1, 4)$ and $(2, 1)$

b) the line through points $(6, 3)$ and $(5, 5)$

c) the line with slope $\frac{1}{3}$ through the point $(0, 5)$

d) the line which is parallel to $y = 2x - 5$ but goes through the point $(1, 7)$

BC-70. Explain, to a student who was absent today, how you would find an equation of the line through two given points.

BC-71. **The Burning Candle** Now you have all the algebraic tools you need to solve the Burning Candle Problem:

Suppose it's your friend's birthday and you want to surprise her by walking into the room carrying a piece of cake with a lighted candle. Could you predict how long before the candle goes out?

In this problem you will use two data points,
(50 seconds, 0.78 grams) and
(3 minutes 50 seconds, 0.57 grams),
to find the slope-intercept form of the equation of the line, and then use the equation to find when the candle would burn out.

THE BURNING CANDLE

a) Use a generic slope triangle to calculate the slope of the line.

b) Follow the third step in BC-66 using one of the candle data points to write an equation of the line in slope-intercept ($y = mx + b$) form, messy numbers and all.

c) What value will y have when the candle burns out? Use that value to replace y and solve the equation you found in part (b) for x.

BC-72. If the slope of line AB is $\frac{5}{9}$, and the coordinates of point B are (18, 6), what are the coordinates of the y-intercept of line AB ?

BC-73. Write an equation for the line passing through each pair of points:

a) S (2, 3) and Y (5, 1)

b) L (1, 7) and V (4, 5)

c) I (–3, 2) and A (1, 6)

BC-74. Use what you know about the slopes of line segments which are on the same line to show that points A, B, and C lie on a line.

A (4, 2)   B (17, 28)   C (7, 8)

BC-75. Solve each of the following equations.

a) $x^2 - 1 = 15$

b) $x^2 - 2x - 8 = 0$

c) $3(x^2 - 1) + 7 = 16$

d) $\sqrt{x^2 - 4} = 16$

BC-76.   Solve each of the following systems of equations.

      a) $y = 3x - 2$
             $y = 4x + 3$

      b) $2x - y = 16$
             $x + y = 14$

      c) $x^2 - y = 9$
             $3x + y = 19$

      d) $2x - y = 16$
             $3x + 4y = 24$

BC-77.   Find the area of a right triangle with a hypotenuse of length 13 centimeters and one leg of length 5 centimeters. What subproblem did you need to solve?

BC-78.   A rectangle is three times as long as it is wide. If the length and width are each decreased by four units, the area is decreased by 176 square units. What are the dimensions of the original rectangle?

BC-79.   **Chapter 9 Summary: Rough Draft** Write your answers to the following questions in rough draft form **on separate sheets of paper**, and be ready to discuss them with your group at the next class meeting. Focus on the **content**, not neatness or appearance, as you write your summary. You will have the chance to revise your work after discussing the rough draft with your group.

      a)   What are the main ideas and skills for this chapter? Give examples to illustrate each one.

      b)   How have you used the idea of ratio throughout this course?

      c)   Suppose you are given two points on a graph. What subproblems would you solve to write an equation of the line through those points?

      d)   What were the most difficult parts of this chapter? List sample problems with your solutions and discuss the hard parts.

      e)   Which problem did you like best and what did you like about it?

9.6 SUMMARY AND REVIEW

BC-80.   **Chapter 9 Summary: Group Discussion** Take out the rough draft summary you
           completed in BC-79. Use some class time to discuss your work; use homework time to revise
           your summaries as needed.

           For each of the three main ideas of the chapter, choose one member of the group to lead a short
           discussion. The discussion leaders should take turns to:

           • explain the problem they chose to illustrate their main idea,

           • explain why they chose that particular problem,

           • tell which problem they liked best and what they liked about it, and

           • tell what they thought were the most difficult parts of this chapter.

           This is your chance to make sure your summary is complete, update your Tool Kits, and work
           together on problems you may not be able to solve yet.

BC-81.   **The Burning Candle Revisited** In BC-71 you solved the Burning Candle Problem
           algebraically. You could also use similar triangles to write an equation that models the
           situation.

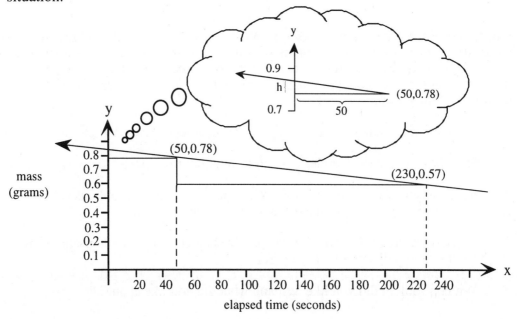

a)   In the picture, two data points,

                    (50 seconds, 0.78 grams) and (3 minutes 50 seconds, 0.57 grams),

       were graphed and two slope triangles were drawn (the smaller triangle is shown enlarged
       in the cloud). Find the altitude, h, of the smaller triangle by using the fact that the two
       slope triangles are similar and writing a ratio equation.

b)   Use the value of h to figure out the y-intercept of the line.

c)   Write an equation of the line in slope-intercept form. Is it the same as you found in
       problem BC-71 ?

BC-82.   Write each of the following expressions as simply as possible.

a) $\dfrac{(x - 1)(x + 2)(x - 3)(x + 4)}{(x + 2)(x + 3)(x + 4)(x + 5)}$

b) $\dfrac{(x^2 - 1)(x^2 - 2x - 3)}{(x - 1)^2(x - 2)(x - 3)}$

BC-83.   Find the exact value of k so that points A, B, and C are on the same line:

$$A\ (1, 0)\qquad B\ (5, k)\qquad C\ (10, 3)$$

a)   Draw a diagram and find the value of k.

b)   Write an equation for the line.

BC-84.   Suppose the slope of line AB is 5, where A is $(-3, -1)$ and B is $(2, n)$. Draw a diagram and find the value of n.

BC-85.   Find an equation of the line …

a)   that has y-intercept 5 and slope 2.

b)   that has slope −3 and goes through the point $(-4, 9)$.

c)   through points $(-1, 2)$ and $(5, -4)$. (Hint: What do you need to do first?)

d)   through points $(-6, 4)$ and $(-2, 1)$.

BC-86.   The perimeter of a certain rectangle is 28 centimeters. The length of the rectangle is four centimeters more than its width. Write an equation to represent this information, find the length and width, and write the ratio of length to width.

BC-87.   **Temperature Conversion Revisited** On a thermometer, 20° Celsius is equal to 68° Fahrenheit. Also, 10° Celsius is equal to 50° Fahrenheit. The conversion from degrees Celsius to degrees Fahrenheit is linear, which means that if the conversion data were plotted as points, the points would lie on a line

    a)   Putting degrees Celsius on the horizontal axis, use the data to plot the points. Then draw a line through the points.

    b)   Use the data to write an equation for the line.

    c)   In part (b) you wrote a formula for converting degrees Celsius to degrees Fahrenheit. Notice that your formula is simply a linear equation. Use your conversion equation to determine the temperature in degrees Fahrenheit for 30° Celsius. Check this with your graph.

    d)   Compare the temperature conversion graph you made in part (a) with the graph you made for problem AP-36. Write one or two sentences to explain how and why they are different.

    e)   Compare the temperature conversion equation you wrote in part (b) with the one you were given in problem AP-36. Write one or two sentences to explain how and why they are different.

BC-88.   **Chapter 9 Summary: Revision** This is the final summary problem for Chapter 9. Using your rough draft from BC-79 and the ideas your group discussed for BC-80, spend time revising and refining your Chapter 9 Summary. Your presentation should be thorough and organized, and should be done on a separate piece of paper.

Algebra Tool Kit: the "what to do when you don't remember what to do" kit

Algebra Tool Kit: the "what to do when you don't remember what to do" kit

Dot Paper

Dot Paper

BC-19. Slopes of Segments

Draw each of the following segments: DE, ED, DF, GH, CF, EF, and EG. Next to each segment, write its slope.

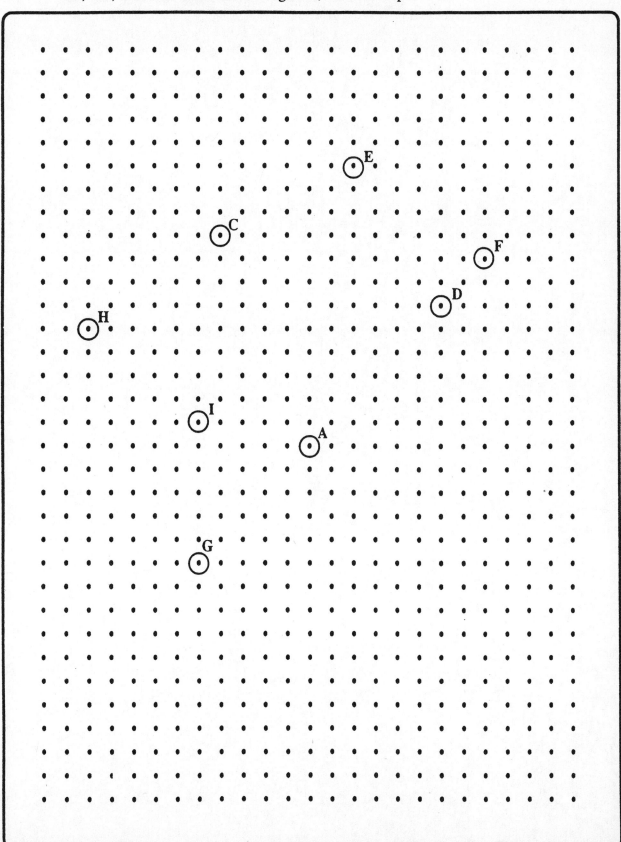

Dot Paper

BC-35. Sketch the graphs and look for similarities and differences.

a)   y = x

y = 2x

y = 4.9x

What happens to the graph of y = x when the coefficient of x is greater than 1; that is, when x is multiplied by a number greater than 1 ?

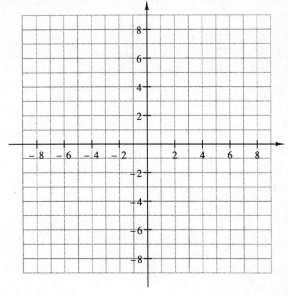

b)   y = x

$y = \frac{1}{2}x$

$y = \frac{1}{4}x$

$y = \frac{1}{6}x$

What happens to the graph of y = x when the coefficient of x is between zero and one?

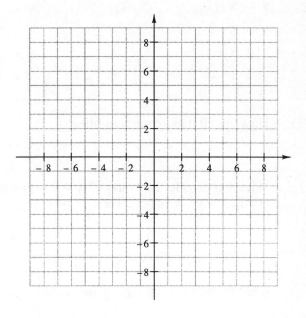

BC-35. continued Sketch the graphs and look for similarities and differences.

c) $y = x$

       $y = \frac{-1}{4} x$

       $y = \frac{-1}{2} x$

       $y = -4x$

       $y = -2x$

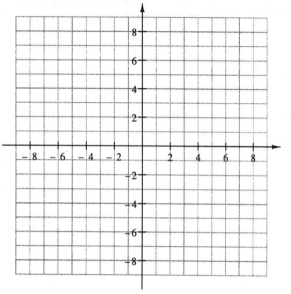

What happens to the graph of $y = x$ when the coefficient of x is negative?

d) $y = x$

       $y = x + 2$

       $y = x - 2$

       $y = x + 4$

       $y = x - 4$

e)   How are all these equations alike?

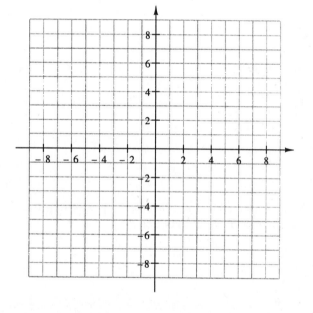

f)   What happens to the graph of $y = x$ when the coefficient of x is negative?

g)   Identify the slope and the y-intercept for each of the following equations.

$y = x$	slope = _____	y-intercept = _____
$y = 0.5x$	slope = _____	y-intercept = _____
$y = 0.5x + 5$	slope = _____	y-intercept = _____
$y = -0.5x$	slope = _____	y-intercept = _____
$y = 0.5x - 5$	slope = _____	y-intercept = _____
$y = -0.5x - 5$	slope = _____	y-intercept = _____

Chapter 10
THE ELECTION POSTER:
More about Quadratic and Linear Equations

CHAPTER 10

THE ELECTION POSTER:
MORE ABOUT QUADRATIC AND LINEAR EQUATIONS

Well, here it is, the final chapter of *Data, Equations and Graphs*! Our goal in this course was to make sense of the important concepts of elementary algebra. In this chapter we tie together many ideas and skills you've learned by focusing on quadratic equations and linear equations. A problem that allows you to use what you've learned about writing and solving quadratic equations is this one about an Election Poster:

Malcolm is running for president of the student body. He orders a poster seven feet wide and five feet high. Irma, Malcolm's campaign manager, decides the poster looks out of proportion and estimates another 75 square feet of area is needed to make it look just right. If each dimension of the poster is increased by the same amount, what are the dimensions of the new poster?

To pull together your skills in graphing data, looking for patterns, and writing linear equations, you'll examine data Ms Speedi's class gathered for a problem similar to the Estimating Book Costs problem you did in Chapter 1.

In this chapter you will have the opportunity to:

- extend your calculator skills for use in solving quadratic equations;

- use the quadratic formula to solve quadratic equations that are not easily solved using the zero product property;

- explore the graphs of various families of quadratics to discover some general properties of parabolas;

- solve a series of subproblems to derive the quadratic formula; and

- use the idea of the line of best fit to connect some important skills and ideas relating graphing data and writing equations of lines.

CHAPTER CONTENTS

10.1 GRAPHS, FACTORING, AND X-INTERCEPTS

You've been solving factorable quadratic equations using the zero product property since Chapter 6. We start off this final chapter of exploring data, equations, and graphs with a series of problems that focus on the solutions of quadratic equations. The problems remind you about multiple solutions to equations, have you practice using the zero product property on easily factorable expressions, and then practice on not-so-easily factorable expressions. The dilemma that appears in EP-4 -- how to find the solutions to a quadratic equation which cannot be factored -- raises the question, "Now what do we do?" This leads to practicing calculator skills to prepare for the answer in the next section.

EP-1.   Find as many solutions as you can for each of the following equations.

   a) $3x = x$

   b) $x^2 = x$

   c) $x^3 = x$

EP-2.   a)   Graph the equation $y = x^2 - 4$ for $-5 \le x \le 5$ and estimate where the graph crosses the x-axis.

   b)   Solve the equation $0 = x^2 - 4$ for x by factoring.

   c)   Write a sentence comparing your solutions to parts (a) and (b). Be sure to note differences, if any, as well as similarities.

EP-3.   a)   Graph the quadratic equation $y = 2x^2 - 7x + 3$ for $-4 \le x \le 4$ and estimate where the graph crosses the x-axis (the x-intercepts).

   b)   Solve the equation $2x^2 - 7x + 3 = 0$ for x.

   c)   Write a sentence or two comparing the solutions to parts (a) and (b).

   d)   How are the solutions to the quadratic equation $2x^2 - 7x + 3 = 0$ related to the x-intercepts of the graph of $y = 2x^2 - 7x + 3$?

EP-4.   a)   Graph the equation $y = x^2 + 6x + 2$ and estimate the x-intercepts.

   b)   Substitute the values you found in part (a) for x into the equation $y = x^2 + 6x + 2$. Solve for y.

   c)   How close to correct were your estimates in part (a) ?

   d)   Is the equation $x^2 + 6x + 2$ factorable? Explain your answer.

   e)   Are there solutions to the equation $0 = x^2 + 6x + 2$? Explain your answer.

EP-5. A calculator is a convenient tool for finding the approximate values of complicated expressions that contain square roots of non-square numbers.

a) Follow each of the three examples below and approximate each value to the nearest hundredth.

Example 1

To calculate $\dfrac{2 - \sqrt{7}}{\sqrt{5}}$, press $[\ \boxed{2}\ \boxed{-}\ \boxed{7}\ \boxed{\sqrt{}}\ ]\ \boxed{\div}\ \boxed{5}\ \boxed{\sqrt{}}\ \boxed{=}$.

Note: If we don't enclose the calculation of the numerator in grouping symbols [ ] before dividing, the result will equal $2 - \dfrac{\sqrt{7}}{\sqrt{5}}$. Does $2 - \dfrac{\sqrt{7}}{\sqrt{5}}$ equal $\dfrac{2 - \sqrt{7}}{\sqrt{5}}$? Explain.

Example 2

To calculate $\sqrt{36 - 4(-4)}$, press $\boxed{36}\ \boxed{-}\ \boxed{4}\ \boxed{x}\ \boxed{4}\ \boxed{\pm}\ \boxed{=}\ \boxed{\sqrt{}}$.

Note: It is critical to press $\boxed{=}$ to complete the calculation under the radical before pressing $\boxed{\sqrt{}}$.

Example 3

To calculate $\dfrac{\sqrt{5} - 2}{3 - \sqrt{7}}$, press $[\ \boxed{5}\ \boxed{\sqrt{}}\ \boxed{-}\ \boxed{2}\ ]\ \boxed{\div}\ [\ \boxed{3}\ \boxed{-}\ \boxed{7}\ \boxed{\sqrt{}}\ ]\ \boxed{=}$.

Note: It is critical to use grouping symbols [ ] in the numerator and denominator to obtain the correct value for the expression.

b) Now use a calculator and what you learned in part (a) to find the approximate value of each of the following expressions to the nearest hundredth.

i) $6 + \sqrt{17}$

ii) $\dfrac{2}{\sqrt{6}}$

iii) $-13 - \sqrt{32}$

iv) $\dfrac{3}{2 - \sqrt{5}}$

EP-6. This problem is a bit more complicated than those in EP-5:

Find an approximate value of $\dfrac{-2 + \sqrt{(-3)^2 - 4(2)(-3)}}{-3}$ on a calculator.

Before you pick up a calculator, list the sub-problems that you will need to solve in order to carry out the calculation.

EP-7.   Use a calculator to find the approximate value of each of the following expressions to the nearest hundredth. Remember to use grouping symbols [ ] when they're needed..

a) $-2 - \sqrt{7}$

b) $\dfrac{8 + \sqrt{17}}{6}$

c) $\dfrac{2}{\sqrt{5}}$

d) $\dfrac{6 - \sqrt{94}}{2}$

EP-8.   Two similar right triangles are cut out of the same piece of flat steel plate. The first triangle has a long leg of 17 centimeters, a short leg of 9 centimeters, and weighs 100 grams. The second triangle has a short leg of 15 centimeters. How much does the second triangle weigh?

EP-9.   Sally has a box full of ping-pong balls numbered from 1 to 53. Charlie reached in and pulled out one ping-pong ball. Find the probability that the number on the ball he chose was ...

a)   even.

b)   less than or equal to 10.

c)   a perfect square (1, 4, 9, 16, …).

d)   a two-digit number.

EP-10.   A rectangle is 2.4 times as high as it is wide and has a diagonal of 52 inches. Find its dimensions.

EP-11.   The two points $(3, -2)$ and $(5, 7)$ determine a line.

a)   Find the slope of the line.

b)   Find an equation of the line.

EP-12.   Use a calculator to find the value of each of the following expressions to the nearest hundredth. Don't forget to use grouping symbols [ ] when they're needed.

a) $\dfrac{-2 - \sqrt{43}}{4}$

b) $\dfrac{7}{6 + \sqrt{8}}$

EP-13.   Find where the graphs of the lines $y = 2x - 7$ and $y = -x + 5$ intersect.

EP-14.   We know that the values $x = 0$ and $x = 1$ satisfy the equation $x^2 - x = 0$. Use a graph of the equation $y = x^2 - x$ to show why no other values of x could also satisfy $x^2 - x = 0$.

EP-15.   A right triangle has an area of 40 square centimeters and its shortest side has length 8 centimeters. Find the length of the hypotenuse.

10.2 THE QUADRATIC FORMULA

Problems such as EP-4 (and even EP-3) suggest that a method other than using factoring and the zero product property is needed for solving quadratic equations. Happily, there is an algebraic tool that accomplishes this task; it's called the quadratic formula. Before introducing the quadratic formula, we familiarize you with the standard form of a quadratic equation.

EP-16.   Use a calculator to find the value of each of the following expressions to the nearest hundredth. Don't forget to use grouping symbols [ ] when they're needed.

Notice the "±" symbol in parts (c) and (d). It provides a way to write two related values at one time. For example, "3 ± 5" is compact way to indicate "3 + 5 or 3 – 5."

a) $8 + \sqrt{8^2 - (4)(5)(3)}$

b) $\dfrac{-2 + \sqrt{2^2 - 4(3)(-7)}}{2(3)}$

c) $\dfrac{-2 \pm \sqrt{16}}{2}$

d) $\dfrac{1 \pm \sqrt{5}}{2}$

EP-17.   **The Standard Form of a Quadratic Equation** Before you can use the quadratic formula to solve a quadratic equation, the equation must be in what we call **standard form**. When in standard form a quadratic equation reads

$$ax^2 + bx + c = 0.$$

The coefficient of x^2 is **a**, the coefficient of x is **b**, and the constant term is **c**. The values of **a**, **b**, and **c** are easily recognized when a quadratic equation is in standard form. For example, the quadratic equation $2x^2 - 13x + 21 = 0$ is in standard form, with a = 2, b = –13, and c = 21.

Identify the values of the coefficients **a** and **b**, and the constant **c**, in each of the following quadratic equations.

a) $3x^2 - 5x + 4 = 0$

b) $x^2 + 9x - 1 = 0$

c) $-\dfrac{1}{2}x^2 + 2x - \dfrac{1}{4} = 0$

d) $-2x^2 + 9x = 0$

e) $0.017x^2 - 0.4x + 20 = 0$

EP-18.   A quadratic equation which is not in standard form must be converted to standard form *before* you can apply the quadratic formula. For example, the quadratic equation

$$3x^2 = 14x - 8$$

is not in standard form. To rewrite it in standard form, we could subtract $14x$ from each side, and add 8 to each side of the equation. Doing this, we get

$$3x^2 - 14x + 8 = 0$$

which **is** in standard form, with $a = 3$, $b = -14$, and $c = 8$.

Rewrite each of the following equations in standard form. In each case. identify the coefficients a and b, and the constant c.

a) $2x + 1 = \frac{-1}{2}x^2$

b) $-6x + 5x^2 = 8$

c) $x(2x + 4) = 7x - 5$

d) $(x - 3)(x + 4) = 7x$

EP-19.   **The Quadratic Formula**   Once a quadratic equation is in the form $ax^2 + bx + c = 0$, you can use the values for a, b, and c to calculate the solutions for the equation. That is, you can find those values of x that make the equation true by using the **quadratic formula**:

$$\text{If } ax^2 + bx + c = 0, \text{ then } x = \frac{-b \pm \sqrt{b^2 - 4ac}}{2a}.$$

This says,

"The values of x which make the quadratic equation $ax^2 + bx + c = 0$ true are equal to the opposite of **b** plus or minus the square root of the quantity, **b** squared minus the product **4ac**, *all* divided by the product **2a**."

Notice that the $\pm$ is *not* a sign change like the key on your calculator. It is an efficient, shorthand way to write the *two* solutions to the equation. The formula actually says:

$$x = \frac{-b + \sqrt{b^2 - 4ac}}{2a} \quad or \quad x = \frac{-b - \sqrt{b^2 - 4ac}}{2a}.$$

Unless the quantity $b^2 - 4ac$ is negative or zero, the graph of $ax^2 + bx + c = 0$ crosses the x-axis at *two points*. For example, the quadratic equation $x^2 + 5x + 3 = 0$ is in standard form with $a = 1, b = 5$, and $c = 3$. When we substitute these values into the quadratic formula we get :

$$x = \frac{-5 \pm \sqrt{5^2 - 4(1)(3)}}{2(1)} = \frac{-5 \pm \sqrt{25-12}}{2} = \frac{-5 \pm \sqrt{13}}{2},$$

which means $x = \frac{-5 + \sqrt{13}}{2}$ <u>or</u> $x = \frac{-5 - \sqrt{13}}{2}$.

These values of x are the EXACT solutions to the equation $x^2 + 5x + 3 = 0$.

[PROBLEM CONTINUED ON NEXT PAGE]

EP-19. continued

We can use our calculators to find approximate values for x:

$$x = \frac{-5 + \sqrt{13}}{2} \approx -0.70$$

or

$$x = \frac{-5 - \sqrt{13}}{2} \approx -4.30$$

Copy the following procedure in your Tool Kit:

How to Use the Quadratic Formula to Solve a Quadratic Equation
Step 1   Put the equation in standard form (zero on one side of the equation).
Step 2   List the numerical values of the coefficients a, b, and c.
Step 3   Write the quadratic formula, <u>even if it's given</u>.
Step 4   Substitute the numerical values for a, b, and c in the quadratic formula.
Step 5   Simplify to get the exact solutions.
Step 6   Use a calculator, if necessary, to get approximate solutions.

EP-20.   **How to Use the Quadratic Formula to Solve a Quadratic Equation**   Here's an example of how to use the six step procedure given in EP-19 to solve a quadratic equation:

Example   Use the quadratic formula to solve the equation $x^2 - 2x = 4$.

Step 1 $x^2 - 2x - 4 = 0$

Step 2 $a = 1$
$b = -2$
$c = -4$

Step 3   If $ax^2 + bx + c = 0$, then $x = \dfrac{-b \pm \sqrt{b^2 - 4ac}}{2a}$.

Step 4 $x = \dfrac{-(-2) \pm \sqrt{(-2)^2 - 4(1)(-4)}}{2(1)}$

Step 5 $x = \dfrac{2 \pm \sqrt{4 + 16}}{2} = \dfrac{2 \pm \sqrt{20}}{2}$

Step 6 $x \approx 3.24$ or $x \approx -1.24$

Solve each of the following quadratic equations by following the six step procedure for using the quadratic formula as shown in the example. Write your answers first in square root form, similar to that in Step 5 of the example, and then in decimal form.

a) $3x^2 - 6x - 45 = 0$ b) $3x^2 + 7x = -2$

c) $(x - 3)(x + 4) = 7x$ d) $8x^2 + 10x + 3 = 0$

EP-21.   a)   Use the quadratic formula to solve the equation $0 = x^2 + 6x + 2$ from EP-4.

         b)   Where does the graph cross the x-axis?

         c)   How close to correct were your original estimates in EP-4 ?

EP-22.   a)   Look again at the polynomial $3x^2 - 6x - 45$ from EP-20(c). Look for a common factor, then use it to rewrite the polynomial.

         b)   Completely factor $3x^2 - 6x - 45$, and then use the zero product property to solve the equation $3x^2 - 6x - 45 = 0$.

         c)   Compare how you solved $3x^2 - 6x - 45 = 0$ in part (b) with what you did in EP-20(c). Think of some advantages and disadvantages of using the quadratic formula. Similarly, think of some advantages and disadvantages of using factoring and the zero product property. Write down a list of your observations.

EP-23.   Tony's driver's license was suspended after his last trip to Los Angeles, so he's making the 305 mile trip from Sacramento by plane. The flight takes only 55 minutes (as opposed to more than five hours by car).

         a)   How many seconds does it take the plane to fly one mile?

         b)   What is the airplane's speed, in terms of miles per hour?

EP-24.   Solve each of the following equations.

         a) $2x + 1 = 7$ b) $2x^2 + 1 = 7$

         c) $5(x - 1) + 3(x + 2) + 2[x + 3(x + 1)] = 25$ d) $\frac{100}{x^2} = 36$

EP-25.   Simplify each of the following expressions.

         a) $5x^2 + 3x - x^2 + 2x + 5 - x$ b) $5(x^2 - 2x) + 2(x - 3)$

         c) $4(x + 2) + 3(x + 2) + 2(2 + x)$ d) $-3(x^2 + 2) + 2(x - 5) - 6(5 - x^2)$

EP-26.   Find the x-intercept, the y-intercept, and the slope of the line for each of the following equations.

         a) $2x - 3y = 7$ b) $y - 4 = 0$

         c) $\frac{2}{x} = \frac{3}{y}$ d) $x + 44 = 0$

EP-27.   A rectangular box has dimensions three centimeters by four centimeters by five centimeters. It has eight vertices. Find as many different distances as you can between pairs of vertices. How many different distances did you find?

EP-28.   A line parallel to $2x + 3y = 4$ goes through the point $(3, 8)$. Find an equation of the line.

10.3 FACTORABILITY AND THE QUADRATIC FORMULA

As dependable as the quadratic formula is for finding solutions to quadratic equations, in some cases it's more convenient to be able to factor the polynomial. In the following problems you'll make a connection between the factorability of a quadratic equation and a special number, $b^2 - 4ac$, the discriminant of the equation $ax^2 + bx + c = 0$.

EP-29.   Use the quadratic formula to solve each of the following equations. Be sure to use all six steps of the procedure.

a) $-3x = -x^2 + 14$

b) $-4x^2 + 8x + 3 = 0$

c) $5 = 6x - x^2$

d) $0.09x^2 - 0.86x + 2 = 0$

e) $3x^2 + 4x = 0$

f) $25x^2 - 49 = 0$

g)   Which of these equations could you have solved by factoring?

EP-30.   Look again at the equations in EP-29 and your solutions to them.

a)   List the values of $\sqrt{b^2 - 4ac}$ for each part of problem EP-29.

b)   Find the equations that had rational numbers (integers or fractions) for their solutions. Compare the values of $\sqrt{b^2 - 4ac}$ for these equations to the values of $\sqrt{b^2 - 4ac}$ for the equations whose solutions were not integers or fractions. What do you notice?

c)   Can $3x^2 + 8x + 5 = 0$ be solved by factoring? Find out by attempting to factor and solve.

d)   Calculate the value of $\sqrt{b^2 - 4ac}$ for the equation $3x^2 + 8x + 5 = 0$. What kind of number do you get? Write one or two sentences that explains the relation of this result to your conclusion in part (b).

EP-31.   **The Discriminant of a Quadratic Equation** The number $b^2 - 4ac$ is called the **discriminant** of the equation $ax^2 + bx + c = 0$.

    a)   Use what you learned in EP-30 to describe the solutions to $ax^2 + bx + c = 0$ when the discriminant is the square of an integer. What if $b^2 - 4ac$ is not a square -- what form will the exact solutions have?

    b)   How could you use the number $b^2 - 4ac$ to tell whether or not the polynomial $ax^2 + bx + c$ is factorable?

EP-32.   Apply your conclusion from problem EP-30 to test if each of the following quadratic equations can be solved by factoring. *Remember to put each equation in standard form first!*

    a) $x^2 + 4x - 5 = 0$ b) $5x = -x^2 + 6$

    c) $x^2 + 7x + 5 = 0$ d) $5x^2 = 6 + 14x$

EP-33.   Solve each of the following equations either by factoring or by using the quadratic formula, whichever method you prefer.

    a) $x^2 + 8x + 5 = 0$

    b) $2x^2 + 5x + 3 = 0$

    c) $x^2 + 7x + 5 = 0$

    d) $x^2 = 10 - 3x$

EP-34.   Mario drew a right triangle with one leg two centimeters longer than the other. The hypotenuse of the triangle was 17 centimeters long.

    a)   Draw a diagram of a right triangle and label the length of each side.

    b)   Write an equation to find the lengths of the legs.

    c)   Find the length of each leg of the triangle Mario drew.

EP-35.   **The Election Poster**   Malcolm is running for president of the student body. He orders a poster seven feet wide and five feet high. Irma, Malcolm's campaign manager, decides the poster looks out of proportion and estimates another 75 square feet of area is needed to make it look just right. If each dimension of the poster is increased by the same amount, what are the dimensions of the new poster?

EP-36.   Malcolm's opponent, Molly Ann, also enlarged her original campaign poster. She started with a 4-foot by 5-foot picture of herself. To enlarge the poster she surrounded the picture with a bright border, which had an area of 50 square feet. If adding the border to the picture increased the width of the original poster and the length of the original poster by the same amounts, what are the dimensions of the new poster?

EP-37.   A square of side length s has the same area as a circle of radius r. Write an equation which expresses r in terms of s.

EP-38.   Use the slope and y-intercept to graph each of the following equations.

a) $y = 2x - 4$ b) $y = -0.5x + 2$

EP-39.   Karen Camero drives straight south from the I-5–Capitol Freeway interchange going 10 miles per hour faster than Farah Fiero, who leaves the same interchange at the same time going straight east on Highway 50. After one hour the two cars are 108 miles apart (as the crow flies). If the highways were actually straight, how fast would each driver be going? Use a solution procedure similar to what you did in problem EP-34.

EP-41.   Find an equation of the line …

a)   with slope $\frac{2}{3}$ passing through the point $(-6, -1)$.

b)   passing through the points $(6, 3)$ and $(5, 5)$.

EP-42.   A certain rectangle has area 50 square meters and its length is five more than twice its width. Find the lengths of the sides of the rectangle.

a)   Draw a diagram and label the length of each side. Think about whether you want to use "x" for the length or for the width.

b)   Write an equation for the area in terms of the width and length.

c)   Find the lengths of each side of the rectangle.

EP-43.   Find all possible solutions to each of the following equations. Explain how you solved each
          equation.

          a) $x^2 = 9$ b) $x^2 = 7$

          c) $(x - 5)^2 = 36$ d) $(x + 3)^2 = 49$

          e) $x^2 + 8 = 51$ f) $(x - 4)^2 + 9 = 12$

EP-44.   Suppose you are given one x^2 tile and 16 small square tiles.

          a)   It is possible to use some rectangular tiles together with the square tiles you are given to
                get a composite rectangle which is a square. Use a diagram to show how this can be
                done.

          b)   Write the area of the composite square you formed in part (a) as a product and as a sum.

EP-45.   Suppose you are given one x^2 tile and six rectangles.

          a)   How many small squares will you need to make a composite rectangle which is a square?

          b)   Sketch the composite square you could form in part (a).

          c)   Write the area of the composite square as a product and as a sum.

          d)   Now you are given one x^2 tile and 10 rectangles, and repeat parts (a), (b) and (c).

10.4 EXPLORING QUADRATIC EQUATIONS AND THEIR GRAPHS

Problems EP-46 through EP-51 are investigations into some general relationships between the equations of parabolas and their graphs. In them you'll be exploring which part of the equation determines the narrowness of the curve and which part determines whether the parabola opens upward or downward. You'll also be moving (translating) parabolas vertically and/or horizontally in a coordinate grid.

EP-46.   Make a large pair of coordinate axes on a sheet of graph paper by placing the origin at the center of the paper. *Neatly* graph each of the following equations *on this one pair of axes* by making tables and assigning values to x from –4 through 4.

   a) $y = x^2$ b) $y = 3x^2$ c) $y = \frac{1}{3}x^2$

           d) $y = -x^2$ e) $y = -\frac{1}{3}x^2$

EP-47   Look at your graphs for problem EP-46. Write a few sentences to describe some patterns you see in the relationship between quadratic equations and their graphs.

EP-48.   Use your conclusions from the previous problem to decide in your groups how to sketch each of the following equations on the same set of axes you used in EP-47. Use a colored pen or pencil to distinguish the graphs.

   a) $y = \frac{1}{2}x^2$

   b) $y = -2x^2$

EP-49.   Make a large graph on a sheet of graph paper by placing the origin at the center of the paper. *Neatly* graph each of the following equations *on this one graph*, by making tables and assigning values for $-4 \leq x \leq 4$.

   a) $y = x^2$ b) $y = x^2 + 3$ c) $y = x^2 - 2$

   d)   Write a few sentences to describe some patterns you see in the relationship between quadratic equations and their graphs.

EP-50.   Make a large graph on a sheet of graph paper by placing the origin at the center of the paper and scaling each axis two squares to each unit. *Neatly* graph each of the following equations, *all on this one graph*, by making tables and assigning values as suggested.

   a) $y = x^2$ for $-4 \leq x \leq 4$

   b) $y = (x - 3)^2$ for $-1 \leq x \leq 7$

   c) $y = (x + 2)^2$ for $-6 \leq x \leq 2$

   d)   Write a few sentences to describe some patterns you see in the graphs of these quadratic equations.

EP-51.   Summarize your observations from problems EP-46 through EP-50 by writing a paragraph that describes some relationships between quadratic equations and their graphs.

EP-52.   a)   Graph the equation $y = 2(x - 1)^2 - 3$.

    b)   Estimate the points where the graph crosses the x-axis.

    c)   Substitute 0 for y and solve for the actual values of the x-intercepts.

EP-53.   a)   Graph the parabola $y = x^2 - 6x$.

    b)   Use your graph from part (a) to graph the equation $y = x^2 - 6x + 3$.

    c)   Suppose you want to find a number c so that the graph of $y = x^2 - 6x + c$ intersects the x-axis at just one point. How many solutions would the equation $x^2 - 6x + c = 0$ have?

    d)   Use the quadratic formula to find such value for c.

EP-54.   Solve each of the following equations for both x and y.

    a) $2x + 3y = 12$ b) $x^2 + 6y = 24$

    c) $\frac{x}{4} - 2y = 10$ d) $x^2 - y^2 = 25$

EP-55.   The point $(3, -7)$ is on a line with slope $\frac{2}{3}$. Find another point on the line.

10.5 USING SUBPROBLEMS TO DERIVE THE QUADRATIC FORMULA (OPTIONAL)

In this section you'll develop the quadratic formula by solving a series of subproblems. Look for patterns as you solve each problem and discuss the patterns you observe with your group.

EP-56.   **Subproblem 1** Solve each of the following equations for y. Each equation has **two** solutions.

a) $y^2 = 25$ b) $y^2 = 19$

c) $y^2 = 45$ d) $y^2 = 16b^2$

e) $y^2 = 13b^2$ f) $y^2 = \frac{36}{49}$ Write the solutions as fractions.

Stop now to discuss a pattern you see developing in parts (a) through (f).

Use the pattern you see to solve parts (g), (h), and (i).

g) $y^2 = 4a^2$

h) $y^2 = \dfrac{bc}{4a^2}$

i) $y^2 = \dfrac{b^2 - 4ac}{4a^2}$

EP-57.   **Subproblem 2: Completing a Square** You know that the expression $x^2 + 6x$ can be represented by tiles:

Notice that the lengths of the sides are both the same, namely $x + 3$, so if you complete the diagram you will have a composite **square.** To make the diagram a complete composite square, you would need to add some small squares. Indeed, if you add 9 small square tiles, the composite square formed would have area $x^2 + 6x + \mathbf{9}$.

What number would you need to add to each of the following expressions to make each one represent the area of a composite square?

a) $x^2 + 10x +$ _____   b) $x^2 - 8x +$ _____

c) $x^2 + 7x +$ _____   d) $x^2 - 5x +$ _____

e) $x^2 + bx +$ _____   f) $x^2 + \dfrac{b}{a}x +$ _____

EP-58.   **Subproblem 3: Dimensions of Composite Squares** From the example in EP-57 you
know $x^2 + 6x + 9 = (x + 3)^2$, so the dimensions of the composite square are $x + 3$ by $x + 3$.

What are the dimensions of each of the completed squares in EP-57 ?

EP-59.   **Subproblem 4: Adding Algebraic Fractions** Write each of the following sums as a
single algebraic fraction. Remember, before you can add any fractions, you need to find a
common denominator.

a) $\dfrac{3}{4a} + \dfrac{5}{a}$ b) $\dfrac{b^2}{4a} + \dfrac{2}{a}$

c) $\dfrac{7}{4a^2} - \dfrac{c}{a}$ d) $\dfrac{b^2}{4a^2} - \dfrac{c}{a}$

EP-60.   **Derivation of the Quadratic Formula** Now we're ready to derive the quadratic formula.
To do this, you will use the last part of each of the subproblems in EP-56 through EP-59 to
solve the equation $ax^2 + bx + c = 0$ for x. Our goal is to end up with

$$x = \frac{-b \pm \sqrt{b^2 - 4ac}}{2a} \ .$$

Fold a piece of lined paper in half vertically, make a crease, then unfold the paper. Copy the
algebraic steps shown below onto the left-hand side of your paper. Write your answer to each
question to the right of the corresponding algebraic step.

We want to solve the equation $ax^2 + bx + c = 0.$

$x^2 + \dfrac{b}{a}x + \dfrac{c}{a} = 0$ *What do we do to get this?*

$x^2 + \dfrac{b}{a}x = -\dfrac{c}{a}$ *What do we do to get this?*

What do we do to get the next step?

$x^2 + \dfrac{b}{a}x + \dfrac{b^2}{4a^2} = \dfrac{b^2}{4a^2} - \dfrac{c}{a}$

[PROBLEM CONTINUED ON NEXT PAGE]

EP-60. continued

Why do you think we chose $\dfrac{b^2}{4a^2}$?

Now we make a major replacement for the whole left side of the equation:

$$\left(x + \frac{b}{2a}\right)^2 = \frac{b^2}{4a^2} - \frac{c}{a}$$ *What subproblem did we use to make this possible?*

This time we replace the right-hand side:

$$\left(x + \frac{b}{2a}\right)^2 = \frac{b^2 - 4ac}{4a^2}$$ *What subproblem shows we can do this?*

What operation do we need to do to get the next result?

$$x + \frac{b}{2a} = \pm \frac{\sqrt{b^2 - 4ac}}{2a}$$ *On what subproblem are we relying?*

$$x = -\frac{b}{2a} \pm \frac{\sqrt{b^2 - 4ac}}{2a}$$ *What do we do to get this result?*

Finally, we get to the long awaited solution:

$$x = \frac{-b \pm \sqrt{b^2 - 4ac}}{2a}$$ *Which subproblem do we use to get this?*

10.6 DATA POINTS AND THE LINE OF BEST FIT

EP-61.   Ms. Speedi just finished grading her algebra class's two most recent quizzes. Each quiz is worth 25 points. Here are the scores for the first twenty students on her class roster:

Student Number	Quiz 11	Quiz 12
1	14	14
2	24	14
3	14	7
4	14	11
5	18	14
6	10	14
7	17	10
8	21	12
9	8	8
10	6	6
11	11	5
12	10	8
13	25	25
14	15	9
15	13	10
16	15	5
17	20	20
18	24	17
19	5	1
20	17	7

a)   Examine all the scores, and then describe in a few sentences how the scores on Quiz 12 seem to compare to the scores on Quiz 11.

b)   Ms. Speedi would like to know more specifically how the two quiz scores for each student compare, and in general, the scores on the two quizzes compare for the class. She could do this by graphing the data. Set up the axes, with scaling and labels, so that the horizontal axis represents scores on Quiz 11 and the vertical axis represents scores on Quiz 12.

      Suppose each student had scored exactly the same on the second quiz as he or she did on the first quiz. Using a red pen or pencil, plot points which would represent this situation and draw a trend line through them. What is the slope of this line?

c)   Students obviously did not score the same on both quizzes. Use a pencil (or blue pen) to plot their actual scores on the coordinate grid. Describe the general trend or pattern of these scores. If you drew a line showing the trend, would the slope be greater than 1 or would the slope be less than 1 ?

d)   Suppose the students had generally done better on Quiz 12 than an Quiz 11. How would the graph then look? On a small grid, make a rough sketch. Would the slope of a trend line for this graph have a slope greater than 1 or less than 1 ?

e)   Save your graph from part (c)! We'll use it again as we continue our analysis of trend lines.

EP-62.   An important skills needed to work with graphs is the ability to write an equation for a given line. Write an equation in the slope-intercept (y = mx + b) form for each of following the graphs. You may need to estimate to the nearest 0.5.

a)

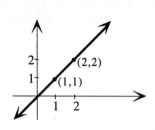

b)

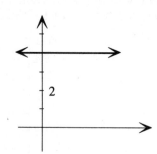

c)

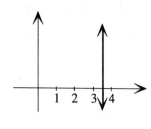

d)

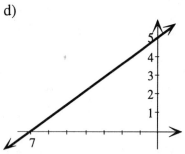

e)

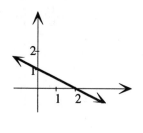

f)

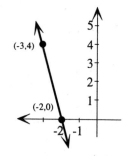

g)

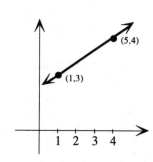

EP-63.*   During World War II, the United States Navy tried to estimate how many German submarines were sunk each month. After the war, the Navy was able to get the actual numbers.

Month	Estimate	Actual number of sunken subs
1	3	3
2	2	2
3	4	6
4	2	3
5	5	4
6	5	3
7	9	11
8	12	9
9	8	10
10	13	16
11	14	13
12	3	5
13	4	6
14	13	19
15	10	15
16	16	15

a)   Make a plot of the estimate x and actual number y of sunken submarines as sixteen ordered pairs, (x, y). For example, the point associated with Month #3 is (4, 6).

b)   Suppose the estimated number of sunken submarines sunk were exactly equal to the actual number of sunken subs. Sketch a line to represent this situation, and then write an equation for the line.

c)   Did the Navy tend to overestimate or underestimate the number of submarines it sank?

d)   Is the line you drew in part (b) a good representation of the pattern (or trend) of the data?

EP-64.   Look again at Ms. Speedi's quiz scores and your graph in EP-61.

a)   Draw a line on the graph which represents the pattern or trend of the scores. This is called a **line of best fit**.

b)   Write an equation for the line you drew in part (a).

c)   Check your work with that of other groups.

d)   Use your equation from part (b) to predict a student's score on Quiz 12 if he or she scored 16 on Quiz 11.

* From *Exploring Data*, by J.M. Landwehr and A.E. Watkins, Dale Seymour Publications.

EP-65.   Julia drives 40 miles per hour faster than Francie. While Francie travels 150 miles, Julia travels 350 miles. Find each person's speed.

EP-66.   a)   Factor the polynomial $4x^2 + 12x + 9$ by drawing a generic rectangle and labeling its dimensions and area(s).

          b)   The polynomial $4x^2 + 12x + 9$ is called a "perfect square" trinomial. Explain why this name is appropriate.

EP-67.   A rectangular garden has a perimeter of 100 meters and a diagonal of 40 meters. Find its dimensions. Solve this problem by writing and solving an equation.

EP-68.   A stick 250 centimeters long is cut into five pieces. The two longer pieces are each 14 centimeters longer than the three shorter ones. How long is each piece? Solve this problem by writing and solving an equation.

EP-69.   The following table gives the length and mass of ten laboratory mice:

length (centimeters)	mass (grams)
21	43
17	38
22	47
16	32
11	19
13	27
15	26
20	40
16	34
15	30

          a)   Examine the data, and then set up coordinate axes, with "length" on the horizontal axis. Scale the axes carefully, and then plot the data as ten ordered pairs (x, y) with x representing length.

          b)   Draw a line of best fit for the data.

          c)   Write an equation for your line of best fit.

          d)   Use your equation from part (c) to predict the mass of a mouse 25 centimeters long.

EP-70. Welded wire fencing is sold in rolls of 100 linear feet. Farmer Fran intends to build a rectangular rabbit pen with an area of 481 square feet. The pen uses exactly one roll of wire, and its length is two feet less than three times its width.

a) Draw a diagram of the pen and label its dimensions.

b) Develop two equations to find the dimensions of the pen, one using area and one using perimeter.

c) Find the dimensions of the pen.

EP-71. If the hundred-foot length of wire in the previous problem weighs 62 pounds, about how much will 240 feet of the same wire weigh?

EP-72. **Chapter 10 Summary: Rough Draft** Five focus ideas for this chapter were mentioned in its Introduction. Look back through the chapter to find where each main idea was developed.

Write your answers to the following questions in rough draft form on separate sheets of paper and be ready to discuss them with your group in class.

a) What did you learn in this chapter that extended or strengthened your understanding of solving quadratic equations? Select one or two problems that illustrates what you now can do. Write complete sentences to describe how you did each of the selected problems. Then tell why you chose the problems that you did.

b) Select a problem you didn't understand before, but now know how to do. Show all your work and a complete solution. Explain why you chose the problem.

c) Select a problem that you liked best or most enjoyed solving. Write the problem and your complete solution. Explain why you chose the problem.

10.7 USING THE MEAN DATA POINT TO FIND A LINE OF BEST FIT (OPTIONAL)

EP-73.   Ms. Speedi calculates the arithmetic mean -- often called the "average" -- of her students'
          scores by adding the scores and then dividing the sum by the number of scores. For example,
          the mean (average) of 12, 15, and 26 is $\dfrac{12 + 15 + 25}{3} = 14.$

          a)   Find the mean (average) of the scores for Ms. Speedi's students for Quiz 11 in EP-61.

          b)   Repeat part (a) for the scores for Quiz 12.

EP-74.   One way to draw a line of best fit is to plot the **mean data point** for a set of data, and then
          "eyeball" a line through it.

          a)   Write the two average quiz scores you found in problem EP-73 for Ms. Speedi's class as
                an ordered pair, (mean for Quiz 11, mean for Quiz 12). This point is the **mean data
                point** for the set of quiz scores.

          b)   On your graph for EP-61, plot the mean data point for the two quizzes. Does it fall on
                your line of best fit? If not, draw a new line of best fit through the mean data point and
                write an equation.

EP-75.   Ms. Speedi's class measured some can lids, paper tubes, rims of bowls,
          and other circular items to try determine a relationship between the
          circumference of a circle and its diameter. Their data are recorded in the
          table below.

diameter (in centimeters)	circumference (in centimeters)
3	10
5	16
10.8	32.8
13	40
10	32.3
6.8	21
4.5	18

          a)   Plot the data, find the mean data point, fit a line, and write an equation for the line.

          b)   In Chapter 8 you used the fact that a circle of radius r units has a circumference of length
                $2\pi r$ units; that is, $C = 2\pi r$. Use this and what you know about circles to write an
                equation for the relationship between the circumference and the diameter of a circle of
                radius r.

          c)   Compare your equation inn part (a) for the line of best fit to the equation in part (b) that
                truly relates circumference and diameter. How closely does line of best fit model the
                actual relationship between the circumference and diameter of a circle?

          d)   Other than measuring more accurately, what else could Ms. Speedi's class do to get a
                more precise description for the circumference-diameter relationship from their data?

EP-76.   Use the mean data point approach to write an equation for a line of best fit for the mouse length and mass problem (EP-69). Show all your work.

EP-77.   Is it always possible to fit a line to a set of data points? Sketch a graph for which you think it would be impossible to fit a line with any accuracy.

EP-78.   Is it possible to fit a <u>curve</u> to data? Sketch a graph for which you think this would be possible. Give an example of an equation for such a curve.

EP-79.   A rectangle has its long side five centimeters longer than its short side and a diagonal five centimeters longer than its long side. Find its dimensions.

EP-80.   Factor each of the following expressions as completely as possible.

a) $x^2 + 4xy + 4y^2$ b) $x^2 + 2y^2 + 3xy$

c) $x^3 + 4x^2 + 4x$ d) $6x^2 + 12 + 18x$

EP-81.   Solve each of the following ratio problems.

a)   Forty-two percent of x is 112. Find x.

b)   Forty-two is x percent of 112. Find x.

c)   Twenty-seven is x percent of 100. Find x.

d)   Twenty-seven percent of 500 is x. Find x.

10.8 SUMMARY AND REVIEW

EP-82.   **Chapter 10 Summary: Group Discussion** Take out the rough draft summary you completed in EP-72. Take turns to:

   • explain the problems you chose to illustrate your understanding of solving quadratic equations;

   • explain why you chose those problems; and

   • explain a problem you didn't understand before, but now can solve, and explain why you chose the problem.

   This is your chance to make sure your summary is complete. You could also update your Tool Kits, if needed, and work together on problems you may not have been able to solve.

EP-83.   A final entry into the race for student body president, Judy, used an enlargement of a 3" by 5" photograph for her campaign poster. She surrounded the enlarged photo with a two-foot border. Including the border, the area of her poster was three times the area of the enlarged photograph. How large was Judy's poster?

EP-84.   Solve each of the following equations.

   a) $x^2 - 2x - 4 = 0$

   b) $x^2 + 6x + 4 = 0$

   c) $3x^2 + 2x - 7 = 0$

   d) $2x^2 - 7x - 4 = 0$

EP-85.   The two solutions to $x^2 - 6x - 8 = 0$ are $3 + \sqrt{17}$ and $3 - \sqrt{17}$. Substitute each of them into the equation to show that they work.

EP-86.   a)   Solve the equation $2x^2 - 7x + 3 = 0$ by factoring.

          b)   Solve the equation $2x^2 - 7x + 3 = 0$ by using the quadratic formula.

          c)   Explain in one or two sentences how you could use the solution from part (b) to factor
                the polynomial $2x^2 - 7x + 3$.

EP-87.   In the previous problem you discussed how to use the solutions obtained from the quadratic
          formula to get factors of a polynomial. Use this idea to factor each of the following
          polynomials.

          a) $6x^2 - 7x - 10 = 0$

          b) $9x^2 + 11x + 2 = 0$

          c) $36x^2 - 37x - 48 = 0$

EP-88.   Use substitution to find x and y if $y = 2x - 7$ and $x^2 + xy = 100$.

EP-89.   Find where the parabolas $y = 2x^2 - 7$ and $y = (x - 4)^2 - 9$ intersect.

EP-90.   Look back at the problems in this chapter where you've graphed data. In these graphs you
          plotted two-variable data. In these cases, there was no direct causal relationship between the
          two variables; that is, one variable did not cause the other. For each of the following
          descriptions, decide if the relationship might be causal, or simply related. Write "causal" or
          "not causal" and state your reasons. Discuss your responses with your group.

          a)   height versus weight of students

          b)   shoe size versus test scores

          c)   exposure to radiation versus cancer rate

          d)   chewing tobacco usage versus incidence of mouth cancer

EP-91.   **Lou's Shoes** At the beginning of the course, in place of the Predicting Book Costs problem Ms. Speedi gave her algebra class a problem about predicting shoe size. The problem was to decide whether or not height was a good predictor of shoe size. Students graphed their shoe size versus their height on coordinate axes and used the data they gathered to form conclusions.

As she cleared off her desk at the end of the term, Ms. Speedi found the data the class had compiled from the graph:

Student	height (inches)	shoe size*
Rob	70	$11^1/_2$
Max	66	9
Jim Bob	68	9
Jamilla	62	7
Karen	62	5
Christina	68	9
Tasha	60	4
Arturo	61	6
Elaine	67	$7^1/_2$
Kelley	65	7

Student	height (inches)	shoe size*
Lawson	76	14
Malcolm	69	12
José	65	$8^1/_2$
Alicia	58	5
Manuel	70	$10^1/_2$
Henry	75	14
Marty	67	10
Lucy	60	$6^1/_2$
Patty	62	6
Lim	71	11

*Sizes were adjusted for the differences between women's and men's shoe sizing.

a)   Plot the data for Ms. Speedi's class. Use height as the independent (x) variable.

b)   Use the mean data point to find a line of best fit for the data.

c)   Write an equation for the line you drew in part (b).

d)   Lou, who is 6'7" tall, was absent the day the data was compiled. Predict his shoe size.

e)   Is there a causal relationship between height and shoe size?

EP-92.   **Chapter 8 Summary: Revision** Use the ideas your group discussed in EP-82 to revise your Chapter 10 Summary. Your presentation should be thorough and organized, and should be done on paper separate from your other work.

EP-93.   **Course Summary Update** Congratulations, your hard work has paid off and you've learned a lot in this course! It's worth spending some time now to reflect on all you've learned and try to pull it all together. Read through the Course Summary Update you wrote at the end of Chapter 8, and then revise it to include important ideas you've learned in the last two chapters. Include copies of homework problems that illustrate these important ideas.

You probably didn't have Ms. Speedi as a teacher in the past, but you probably had a favorite teacher, in math or another subject. Write a brief note to that favorite teacher to tell her/him in general terms what you've learned from working through *Data, Equations, and Graphs* this term. Be sure to describe your greatest accomplishment in the course.

Algebra Tool Kit: the "what to do when you don't remember what to do" kit

Algebra Tool Kit: the "what to do when you don't remember what to do" kit